Sustainable Futures and Green New Deals

Zaheer Allam · Ali Cheshmehzangi

Sustainable Futures and Green New Deals

palgrave
macmillan

Zaheer Allam
Institute of Sustainable Futures
University of Technology Sydney
Sydney, NSW, Australia

Urban Transformations Research Centre (UTRC)
Western Sydney University
Parramatta, Australia

Ali Cheshmehzangi
The University of Queensland
Queensland, Australia

ISBN 978-3-031-63641-7 ISBN 978-3-031-63642-4 (eBook)
https://doi.org/10.1007/978-3-031-63642-4

This Palgrave Macmillan imprint is published by the registered company Springer Nature Switzerland AG
The registered company address is: Gewerbestrasse 11, 6330 Cham, Switzerland

We dedicate this book to our families who have been incredibly supportive and patient in our busy (and writing) times. They are the backbone of our every success.

We also devote this to worldwide green activists and the progressive youth behind sustainability innovation, pathways, and directions.

Acknowledgements

As always, we would like to acknowledge and value the hard work of our collaborators in multiple locations. We respect their devotion, desire, and diligence in supporting our work by providing materials, case study reports, and policy documents. The assistance offered is beyond appreciation, especially towards addressing research development and trajectory of manifestations. Portraying low-key ubiquitous suggestions leads to tragic losses, educational negligence, and ineffective creations. Thus, we hope this book adds value to academic research, scientific knowledge, practice, and policy development in a different way.

About This Book

This academic book critically examines the concept of sustainable futures and its interdependence with the policy framework of Green New Deals. We emphasize the inherent link between the two concepts by highlighting how Green New Deals can pave the way for a sustainable and just future in key areas such as environmental policy, economic development, and social equity. Taking a multidisciplinary approach, the book draws on insights from fields such as environmental studies, sustainable development, political ecology, and climate policy to explore the intricacies of these interconnections. The revised description underscores the importance of understanding the Green New Deal's historical and political contexts, as well as its implications for sustainable futures. The book delves into the theoretical perspectives that shape our comprehension of sustainable futures, the diverse strategies employed by countries and regions in implementing Green New Deals, the role of technological innovation and social movements in fostering sustainable transitions, and the political dynamics surrounding the Green New Deal.

Through an analysis of the Green New Deal's role in sustainable development, climate policy, and environmental justice, the book provides a necessary and timely contribution to ongoing debates. The chapters offer detailed explorations of the interrelationships between these key areas, emphasizing how the Green New Deal can serve as a transformative policy framework for achieving sustainable and equitable futures.

"Sustainable Futures and Green New Deals" is intended for scholars, policymakers, activists, and students interested in understanding the complex and interconnected issues surrounding sustainable development, climate policy, and environmental justice.

Dr. Zaheer Allam
Institute of Sustainable Futures
University of Technology Sydney
Sydney, Australia

Charles Telfair Campus
Curtin University
Terre-Rouge, Mauritius

Prof. Ali Cheshmehzangi
School of Architecture, Design
and Planning
The University of Queensland
Brisbane, Australia

Hiroshima University
Hiroshima, Japan

Contents

About the Authors

Dr. Zaheer Allam possesses a diverse educational background that includes a Ph.D. in Humanities, a Master of Arts in Political Economy, a Master of Sustainable Futures, an MBA, and a Bachelor of Applied Sciences in Architecture, earned from prestigious universities in Australia and the UK. Being Mauritius-based, Dr. Allam served as the first Chairperson of the National Youth Environment Council (NYEC) under the Prime Minister's office, where he advised the Prime Minister on matters of Youth and Environment, and where he currently sits on the board of directors for the Mauritius Renewable Energy Agency (MARENA). Besides being a consultant for UNESCO, he also sits on a high-level committee for UN HABITAT to drive international Guidelines on People-Centered Smart Cities. In his capacity as an independent consultant, Dr. Allam spearheads innovative projects in urban regeneration, smart cities, and urban development, while integrating technology-driven strategies in both cultural and corporate contexts. As an Honorary Fellow of Deakin University (Australia) and Hiroshima University (Japan), and a Research Associate of Sorbonne University (France), he collaborates with interdisciplinary researchers to publish groundbreaking work in leading academic outlets and actively participates in global dialogues through regional organizations.

As the African representative for the International Society of Biourbanism (ISB), Advisory Circle Member of the International Federation of Landscape Architects (IFLA), and a member of numerous international organizations, Dr. Allam's impact on the global stage is centred on policy and advocacy. Honoured by the President of Mauritius with the rank of 'Officer of the Order of the Star and the Key of the Indian Ocean (OSK), Dr. Allam has also received numerous accolades and recognition as a top global scientist, with works featured in international outlets such as The Washington Post, Forbes, NATURE, UNESCO, UN HABITAT, and many others. His body of work includes over 175 scientific publications and 13 books that explore the future of smart and sustainable cities.

Prof. Ali Cheshmehzangi is the World's top 2% field leader, recognized by Stanford University. He is Professor and Head of the School of Architecture, Design, and Planning (ADP) at The University of Queensland, Brisbane, Australia. Previously, he held several key leadership roles and university management positions of various levels, such as Head of Innovation Centers/Labs, Head of Department, Director of Research Centers/Groups, etc. More recently, he led several strategic initiatives on international communications and global partnerships between China and the UK. He has collaborated and successfully completed projects with the Asian Development Bank (ADB), National Development and Reform Commission (NDRC), Building Research Establishment (BRE), Guangzhou Urban Innovation Institute, UN Habitat, and many other major international organizations. His research focuses on urban sustainability and city transitions, while exploring climate resilience, smart resilience, and integrated planning/design innovations.

Moreover, Ali has a wealth of experience, expertise, and a strong commitment to innovation and excellence in the field of architecture, design, and planning. Spanning over two decades, he has made significant contributions to the academic and professional communities, with a focus on sustainable and environmentally conscious design. He has a passion for sustainability, urban resilience, and cutting-edge (urban) planning and design thinking. He was a Specially-appointed Professor and is now a Research Associate at Hiroshima University, Japan. So far, Ali has published over 450 journal papers, articles, conference papers, book chapters, and reports. He has 25 other academic books focused on cities and sustainable development research areas. Three of his books have received several international, national, and provincial/municipal awards. He is also Editor-in-Chief of Springer's Urban Sustainability Book Series, which was launched in February 2022.

List of Figures

List of Tables

Chapter 1

CHAPTER 1

Sustainable Futures and Green New Deals: Overview and Policy Directions

Abstract This chapter addresses the critical challenges of climate change and the necessary policy directions for sustainable futures through Green New Deals (GNDs). Climate change is causing severe health, economic, and environmental issues, including heat-related illnesses, food insecurity, droughts, and rising sea levels. Current efforts to meet the Paris Agreement goals are falling short, with potential temperature rises of 3 °C to 3.5 °C by 2100. Emphasizing the importance of a global approach to GNDs, it highlights their potential to address social, economic, and environmental injustices. The historical and political contexts reveal the evolution of GNDs from the 2006 concept in the United States to varied adaptations worldwide. Key policy proposals focus on the need for international cooperation, just transition strategies, effective engagement across ideological divides, sustainable financing models, and robust progress measurement frameworks. While GNDs offer transformative potential, achieving their goals requires overcoming significant political, economic, logistical, and technological challenges.

Keywords Climate change · Green New Deals (GNDs) · Sustainable futures · Paris Agreement · Environmental justice · Renewable energy · Just transition · International cooperation

Z. Allam and A. Cheshmehzangi, *Sustainable Futures and Green New Deals*, https://doi.org/10.1007/978-3-031-63642-4_1

1 Introduction

The increasing impacts and challenges of climate change are more apparent and have been widely documented, with consequences affecting every aspect of the planet and its inhabitants. For instance, there are now increased cases of heat-related health problems in different parts of the world, diverse economic challenges (such as food insecurity), and a myriad of environmental challenges (Cheshmehzangi & Dawodu, 2019). A case in point is the increased precipitation, leading to flashfloods, like recently witnessed in Western Europe, where over 200 deaths were reported in July 2021 alone (Davies, 2021). Climate change is also credited for the increasing incidences of drought, especially in sub-Saharan Africa with affected regions experiencing food insecurity, increased cases of water shortage, and numerous challenges amongst pastoralists as vast lands experience desertification (Chiang et al., 2021; Dofonsou et al., 2023; Inman et al., 2020). As a result, approximately 70%–80% of people get displaced during the drought season compared to a non-drought period (which has now become rare), thus, compounding human and animal suffering (Ceola et al., 2023). In the coastal regions, the visible rise in water sea levels is becoming very real, threatening some coastal regions with submersion (Roy et al., 2023). Thus, we see significant environmental threats and complications occurring globally.

According to the latest report by the IPCC (2023), it is possible that the climate goals set in a number of global accords like the Paris Agreement, aimed at ensuring that global temperatures are kept below 2 °C, preferably 1.5 °C will not materialize, affirming similar findings documented by the same body in 2018 and 2021 respectively (Intergovernmental Panel on Climate Change, 2018; IPCC, 2021). According to the report, there are possibilities that the temperatures will have risen by between 3 °C and 3.5 °C by 2100 (Cheshmehzangi, 2020; Cheshmehzangi & Chen, 2021). This is reinforced in the Nationally Determined Contributions (NDC) Synthesis report, published on 17 September 2021, outlining that even though the global rate of carbon emissions was decreasing due to economic slow-down due to impacts of COVID-19 (Allam & Jones, 2021; Allam et al., 2021a, 2022a), it was expected to peak by 2030—with a 16% comparative rate of increase to 2010 levels, suggesting an increase in temperature rise of about 2.7 °C; hence, demanding further efforts by countries to achieve the Paris Agreement targets (UNFCCC, 2021). The prediction is still alive, as noted in the 2023 Annual Global Carbon Budget report that predicted a 1.1% (to reach almost 40 billion metric tonnes) (Carbon Brief, 2023) emissions increase from the 2022 reported levels (36.8 billion metric tonnes) (IEA, 2023), which were also higher than the 2021 estimates.

The Paris Agreement, as noted in the UN Climate Change report (United Nations Framework Convention on Climate Change, 2015), required that member parties commit to meet at least a 45% reduction on their emissions by 2030, but it was found that commitments could only be achieved if each party doubled their efforts. The scenario was further complicated by the emergence of COVID-19, which Shan et al. (2021) posited would prompt a 16.4% global increase in emissions from 2020 depending on economic stimulus governments advance in their economies. On this, the UNEP (2021) reported that only 18% of all the global recovery spending had some 'Green' elements. With those facts, it is worthwhile noting that temperatures above 1.5 °C will translate to devastating climate events like heatwaves, flooding, desertification, acidification of water sources and tropical cyclones among others. This may be disastrous for many regions of the world, and specially Small Island Developing States (SIDS) (Allam, 2020a; Allam et al., 2020a, 2020d; Zaheer and Society, 2020), where cases of forced climate migration are expected to increase with seawater level rise, with others losing a considerable amount of habitable spaces, fishing ground, tourism attraction sites, and essential infrastructures, leaving little options other than to migrate (UNEP, 2014).

Failing to meet the agreed targets in the Paris Agreement will consequently impact the achievement of other agreements, like the Sustainable Development Goals (United Nations Development Programme (UNDP), 2015) and that of other COPs that were reached earlier. This however will be a result of human inducement, especially post-COVID-19 pandemic, where economies have been found to revert to fossil fuel-intensive industries in their effort to revive the economies (Allam et al., 2021a; Le Quéré et al., 2021). Such has also been compounded by the ongoing political turmoil such as the Russia-Ukraine war (Rawtani et al., 2022) and the Israel-Palestine conflicts (Feitelson et al., 2012), which have had a huge toll on global sustainability efforts. On this, Shan et al. (2021) note that if fossil fuel intensive activities continue to prevail, the aftermath will be a chaotic global environment that will not only be unsustainable for the current generation but also make sustainable development more difficult for future generations. As expressed by The European Commission

(2020), there will be a need for deep rethinking and reversal in policies to ensure global extraction of resources, production of products and their consumption, and their ultimate disposal align with the sustainability agenda.

One solution that emerged in 2019, and has been gaining traction in the past three years, is the Green New Deal which was first introduced in America after years of negotiations (Congress, 2019) and was subsequently adopted in varying forms by numerous countries, including country blocks like the European Union (European Commission, 2019). This is further explored in the next section.

2 Historical and Political Contexts

The concept of Green New Deals can be argued to be an emerging, broad, and transformative approach towards addressing a series of issues such as increasing social inequality, environmental degradation, and others impacting modern societies. It thus aims to address climate concerns in different geographies, while attempting to solve social and economic inequalities (Walker, 2022). According to Conte (2019), the idea of the Green New Deal was conceived in 2006 in the United States, under the care of a taskforce that had been formed to deliberate and propose solutions on how different sectors engage in a sustainable transition as per the concept of the 'Global Greens'. This is interesting to note that the emergence of associated terms such as 'Green Economy' and 'Green Growth' in 1989 and 2005 respectively (UN, 2021), led to foundational precepts for the Green New Deal. However, the proposed GND, as noted above and highlighted by Mastini et al. (2021), goes beyond economic growth in relative terms and aims to address other aspects like social, economic and environmental justice. In fact, the concept of 'growth' should not be a precursor for the GND, as expressed by Mastini et al. (2021), as degrowth policies also need to be part of the 'GND' narrative. This is true noting that some green growth pursuits also have the potential to violate absolute planetary boundaries rather than solely improving environmental capital, or at least in the preservation of existing environmental resources (Allam et al., 2022b, 2022c; Calhoun & Fong, 2022).

The main inspiration for the Green New Deal, including the name was borrowed to mirror President Franklin D. Roosevelt's approach of helping the United States recover from the Great Depression by instating a total makeover of his government; or the 'New Deal' as it was called

(Calhoun & Fong, 2022). Activism on climate action began in the 1970s after the oil crisis, but progressive actions were only arguably taken in 2006 (Conte, 2019). The agitation for climate change, gained more support from the political class, where some politicians with Jill Stein; a presidential candidate in 2012 and 2016 under her Green Party banked on the climate change debate as a campaign tool for the presidential position (Atkin, 2019). However, the main breakthrough of the political class came in 2018 after a youthful group that was protesting the government's inaction against climate change was joined by Rep. Alexandria Ocasio-Cortez and Senator ED Markey (Congress, 2019), who formulated the groundwork for what became a joint resolution passed by congress in 2019. The resolution mandated the Federal Government to create a Green New Deal that would address a raft of issues, including touching on the foundational challenges of climate, economic, and social justice.

On climate, the Green New Deal (GND) as passed by Congress targeted to have the country (United States) achieve net-zero Greenhouse gas emissions by 2050, by adopting transition mechanisms that would not be injurious to communities or workers. The achievement of reduced emissions would be championed by ensuring the country transitions 100% to the use of alternative renewable energies in all sectors, through deep transformation of the transport sector, ensuring electric vehicles and mass transit systems were the main mode of transport. Since the objectives of the GND match the calls for climate action (Bloomfield & Steward, 2020), there seems to have been an increase in popularity, not just in the United States, towards the model. Indeed, an analysis, represented through a line diagram in Fig. 1, of the popularity of the term 'Green New Deal' for the period of 2008 to 2023 showcases a peak in 2019, aligning with the time of the GND passing in Congress in the United States. Additionally, it is observed (Fig. 2) that the first five geographies where interest by regions is peaking are: Guatemala, South Korea, Singapore, the United States, and New Zealand respectively.

Since the adoption of the Green New Deal in the U.S. Congress in 2019, there have been a substantial number of other countries that have adopted similar approaches under different names, as shown in Table 1.

Though they vary from each other, they all seem to have a convergence of intentions, especially regarding the need to reduce emissions and concentrate on attaining net-zero emissions in the future while achieving equity and inclusivity in the socio-economic fabric. In the table, China

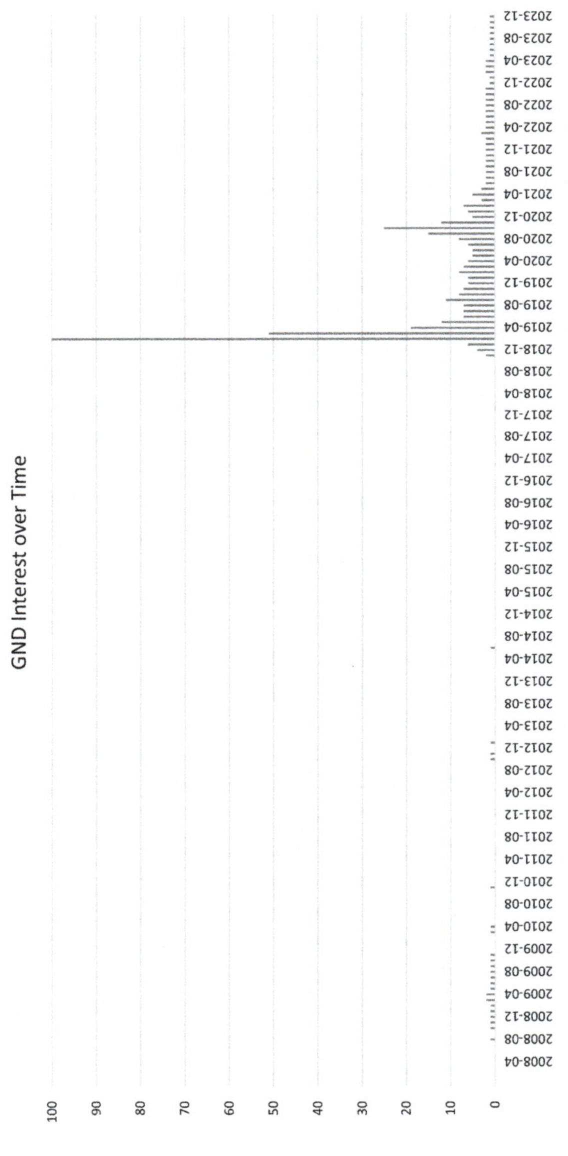

Fig. 1 Web popularity of the term 'Green New Deal' from 2008 to 2023, with popularity ranging from 0 to 100. Sourced from Google Trends (2023)

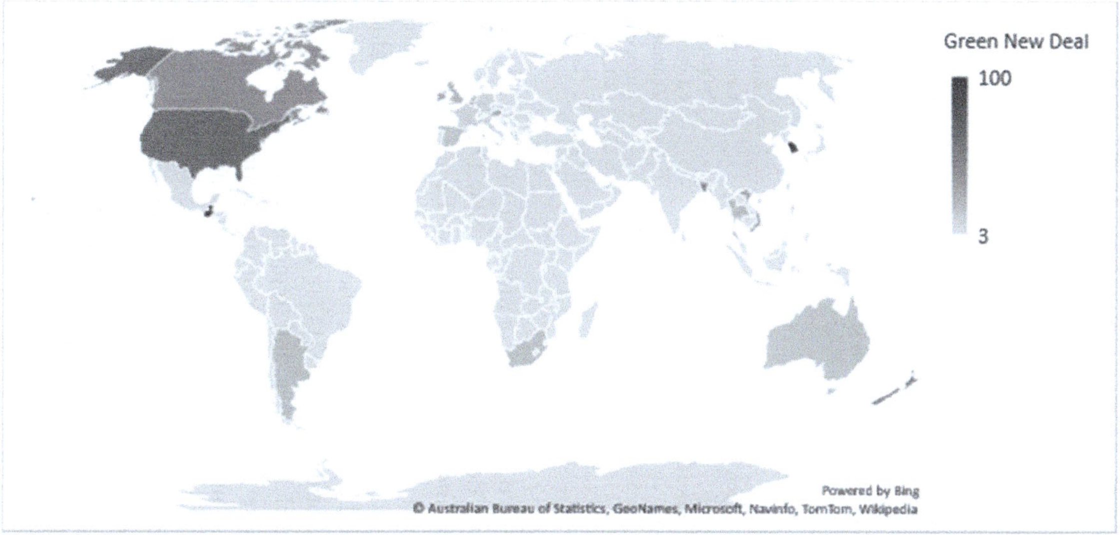

Fig. 2 Web Popularity of the term 'Green New Deal' (2008 to 2021)—Interest by Region. The popularity ranges from 3 to 100, where 100 is the most popular. Sourced from Google Trends (2021)

Table 1 Mapping Green New Deal national policy proposals and name variations

Country	*Name of policy*	*Year proposed*	*Source*
Canada	Pact for a Green New Deal	2019	(MacArthur et al., 2020)
United Kingdom	Green Recovery Act	2019	London Assembly (2021)
South Korea	Green New Deal	2020	Chowdhury (2021)
The European Union	European Green New Deal	2019	European Commission (2019)
China	Ecological Civilization	2020	Paszak (2020)
United States of America	Green New Deal	2019	Congress (2019)
Mexico	Green New Deal	2020	Moreno-Brid and Gallagher (2020)
Singapore	Green Plan	2021	Government of Singapore (2021)
New Zealand	Aotearoa New Zealand Green New Deal	2019	Government of New Zealand (2022)
EU	Fit for 55	2021	European Council (2021)

and India are included, but their commitment to the reduction of emissions is not expressly defined in their Green New Deal agendas (Donato, 2020), but notable and practical actions have been taken. These include increased efforts and investments in renewable energy (Chabaud et al., 2023), adoption of technologies in their manufacturing industries, and introduction of fiscal mechanisms like the use of emission trading schemes (ETS) (Chatzky & Siripurapu, 2021) are evident, especially in China.

Besides individual countries and regions pursuing the GND, it is interesting that there are also notable organizations that support Green New Deal policies, pushing sectoral and thematic agendas on sustainability transitions (Allam et al., 2021b). Those are outlined in Table 2. Most of those are domiciled in countries and regions like the United States and Europe that have already pioneered in forming models of their GND. However, even in countries like Australia where there is no formal announcement of the GND program, from the federal government, some States within the republic are seen to actively advocate for its acknowledgement and the embracement of deeply sustainable transitions (Gunia, 2021). However, there appears to be clear disagreement on what constitutes a 'green deal', especially in view of criticisms of an AUD 2 billion renewable energy project between the New South Wales (NSW) government and the federal government (Visentin, 2020).

Despite the GND emerging as among the most potent and powerful approaches to tackling the interconnected challenges spurred by climate change, increasing social inequalities and economic transformation, it will require concerted effort for them to be achieved. A starting point could be the formulation of a universally acceptable GND variant, which could assimilate ideas, approaches, and inputs from diverse quarters, rather than each country maintaining its own. A practical example is the 'Fit for 55' GND concept proposed for the European countries, which focuses on the entire region (European Council, 2021). But, even as a common GND is pursued, there will be a need to address concerns on issues like feasibility, cost, and economic disruptions such a move would elicit, especially on economies in the global south, the marginalized communities, and people within and outside of different economies. As such, as steps towards the

Table 2 Mapping organizational adoption of Green New Deals

Organizations	*Year*	*Source*
The Climate Mobilization	2017	The Climate Mobilization (2017)
The European Green Party	2006	Green Party (2021)
The Green-European Free Alliance	2011	The Greens/EFA (2011)
The Democracy in Europe Movement 2025	2019	DiEM25 (2019)
Green Party of the United States	2006	Green Party of the United States (2006)
Heinrich Böll Foundation	2009	French et al. (2009)
League of Conservation Voters	2019	League of Conservation Voters (LCV) (2019)
The New Economic Foundation	2008	New Economics Foundation (2008)
Open Democracy	2018	Robinson (2018)
The United Nations Environmental Programme	2009	UNEP (2009)
The Global Marshall Plan Initiative	2020	Saha et al. (2020)
The United Nations Economic and Social Commission for Asia and the Pacific	2012	Ministry of Environment (2020)

GND are being taken, it will be critical for different countries and organizations to learn from each other, gain valuable insights, and create more opportunities worth for a more just and sustainable world.

3 Key Policy Proposals

The concept of GND is a timely idea that can help address the looming interconnected challenges of climate change, social inequality, and economic transformation in different parts of the world. However, as has been discussed in the previous section, it would be distasteful for this concept to succeed at a global scale if each country were to pursue its own model without regard to what its peers are doing on the same matter. Bradshaw et al. (2021) note that while the concept is gaining attention, leading to rapid acceptability and adoption in diverse geographies and quarters—specifically with civil societies pressuring legislative makers—underlining evidence of diversification in terms of the content advanced in the different countries or regional specific models. This has led to confusion on the specific underlying principles that the varied models aim to address thus, pointing to the need for countries to first address underlying market failures, such that the GND is not seen as a knee-jerk reaction to failures that could be sorted using other means. In fact, even in country-specific models, like the Green New Deal in the United States, there is still confusion noted from leaders from different political divides (Democrats Vs. Republicans), as to who is against the deal—notably being mostly Republicans, who perceive some of the issues, like social justice, as being unrealistic; hence being dismissive of the proposal (Bradshaw et al., 2021).

While the variations may be deemed to represent interests that different countries have in their pursuit and conviction of the best way to address climate change, rising oil prices, and unsustainable energy consumption, there is a need for some uniformity to address common goals and principles. This is affirmed when a critical consideration of models like the EU's Green Deal will be expanded beyond the EU as explained by Dartford (2021). While the intention is deemed as sincerely aiming at ensuring that Europe's pursuit of climate action is achieved, it may have deeper implications as the region's trading partners may not have any contextual plans for GND.

The resulting implications from both the trading partners and the EU may be hostile to the institution of stringent measures that may somehow disrupt the existing trading relationship. In the case of countries in the global south, that in some ways rely on their counterparts in the global north, they may be inevitably forced to agree to some measures, which sometimes may be incoherent with their domestic policies. On this, Táíwò (2019) argues that some of the policies engendered in the New Green Deal and other models, if eventually formalized to become guiding policies, have the potential to increase inequalities; akin to what scholars coin as being 'climate colonialism' (Doreen, 2014). Such would not augur well with the underlying principle of 'a just transition' to a sustainable future advanced by the United Nations (United Nations, 2022) which, at the bare minimum requires addressing existing social inequalities and ensuring that everyone benefits from the green economy.

While examples would vary in accordance with contextual needs and capacities, there are notable needs by countries for land and financial resources to develop renewable energy plants as alternatives to fossil fuel plants. This is further represented in the NDC Synthesis Report (UNFCCC, 2021), stating the popularity of Renewable Energy projects, as a means to curb climate change, amounts to 84% in countries around the world. While this measure is welcomed globally, most economies in the global south may have sufficient land, without the associated financial capacities to support nationwide projects, and thus often rely on debt financing, grants and other such sources (Allam & Jones, 2019a, 2019b; Allam et al., 2020c). This conventional approach could force nations to secure more debts, increasing an unsustainable dependency on developed economies, leading to the arguments of 'neo' and 'climate' colonialism (Allam et al., 2020b). In this case therefore, economies in the global south, just like is the case with their counterparts in the north should be allowed to pursue and craft their own sustainability plans, addressing their unique challenges, as well as allowing them to align with their commitments as per the Paris Agreement as well as SDGs; hence calling towards contextualized models for Green New Deals. Alternatively, all the players at the global scale could agree to forge a common understanding on climate financing, that does not discriminate against anyone. For instance, there could be an agreement for green industrialization and diversification, where those with green resources like land, natural gas, hydropower, and others could present them, while those with financial capabilities could invest in such places. This way, it would be possible for each partner to participate in climate change mitigation as equals (Kozul-Wright, 2023).

While pursuing proposals for new models, that would somehow apply conventionally; hence, acceptability in countries and by different organizations, it will be paramount to understand the underlying issues pertaining to GND in terms of how it emerged, its evolution and trends. This is particularly important at a time when GNDs are politically equated as an 'ecology-centered' economic stimulus program, which can help significant political attention (Chohan, 2019). This is key, especially when GNDs can be viewed as a strong narrative to help economic growth, hence posing as a self-funding mechanism to the crucial question of how to finance transitions. From this perspective, proponents of GNDs are largely unaligned with those of degrowth who argue growth makes it more difficult to accomplish emissions reductions (Mastini et al., 2021). This interestingly builds an alignment of agendas on GNDs, between both the environmental and political classes, with the agenda of economic growth as a mutual ground. Additionally, funding transitions across multiple sectors are required, and in this regard, numerous innovations need to be fostered, which can also lead to white space opportunities for corporates. This is leading to emerging Green funds, aligning with the underlying GND theme, applicable at both national and regional levels, and which could provide some reprieve to global south economies without the capacity to fully finance GND agendas.

To align with this constant shift in global policies, including in view of the current COVID-19 pandemic (at the time of writing), there is a need for an updated review of literature on GNDs and associated strategies for Green Growth, as new economic challenges present itself, causing larger inequalities and posing a threat to long-term sustainability agendas (Allam, 2020b; Allam et al., 2021a). In view of this, a macroscopic perspective is required on the larger themes, providing an overview of its evolution over time, including during the pandemic period, to better understand past, current, and future directors. This paper thus, through a bibliometric analysis, undertakes to advance our knowledge of GNDs to present findings that can help countries seeking to associate with GNDs would have substantial information, leading to the potentiality of crafting models with the potential to address local priorities and challenges.

3.1 *Bone of Contention*

The feasibility of GND is contingent upon navigating complex political, economic, logistical, and technological challenges (IPCC, 2021).

But as was put forth by Ocasio-Cortez, the long-term economic benefits outweigh the upfront investments; hence, a worthy course to pursue. Such calls have been taken in different jurisdictions as described in the previous section. Among the feasible routes that are gaining popularity as countries seek to embrace the GND concept regarding addressing the issue of climate change is the adoption of renewable energies. For instance, in Denmark, significant strides in the adoption of renewable energy have been made with wind energy accounting for over 50% of its electricity consumption (International Renewable Energy Agency, 2020). Such is expected to relieve the citizens of some financial burdens, which could be diverted to social aspects. It is worth noting that besides Denmark, there are other numerous countries that have set sights on renewable energy, with resources and relevant policy interventions being put in place. It is anticipated that the huge investments being made in renewable energies would contribute greatly to creating millions of jobs; hence, also help in addressing social injustices and inequality (Allam et al., 2023; Cheshmehzangi et al., 2024). However, scaling up renewable energy infrastructure and implementing energy-efficient measures require substantial investments and coordinated efforts. Critics argue that achieving ambitious emissions reduction targets (even with a focus on renewable energies) and economic transformation within short timeframes is unrealistic and financially unsustainable, especially noting that estimates for the cost of various GND proposals vary widely, ranging from trillions to tens of trillions of dollars (Fischer & Jacobsen, 2021). While there is the potential for cost reductions through technological advancements, public–private partnerships, and efficient policy design intentional interventions cannot be overlooked.

The transition to a green economy may indeed involve disruptions to certain industries, which could have negative consequences if proactive policies to mitigate those impacts on workers, investors, and other stakeholders are not crafted. However, there are some front-runners who could be emulated regarding such policies. For example, Spain's Just Transition Strategy which includes measures to support workers in coal-dependent regions through retraining and job placement programs is a good example (Government of Spain, 2020). Another such is the U.S.' Clean Energy Corps proposal which aims to create 250,000 jobs for young people in environmental conservation and clean energy. The policies need not be complex, but as simple as what has been done in India with its determination to expand its solar energy industry thus creating jobs for rural communities while reducing carbon emissions (The World Bank, 2020). It could also entail investments in energy-efficient buildings and transportation infrastructure to generate employment across various sectors, contributing to economic resilience and social equity (International Renewable Energy Agency, 2020).

A critical review of the economic modelling studies on GND indicates that investments in clean energy and infrastructure can stimulate economic growth and create millions of jobs. For instance, a report by the Economic Policy Institute estimates that investing $2 trillion in clean energy over ten years could create between 6.9 and 12.9 million jobs annually (Scott et al., 2020). Moreover, research suggests that the transition to a green economy can enhance productivity, reduce healthcare costs, and improve overall well-being (Zhou et al., 2022).

The feasibility and cost of GND, plus potential benefits are however premised on the political goodwill at the local, national, and global levels. It has been established that political polarization poses a significant barrier to climate action. For instance, the full brown war between Russia and Ukraine, plus that between Israel and Palestine has been argued to roll back the benefits already made by the adoption of GND in different countries. A case in point is the oil crisis, plus the supply chain disruption due to the reactionary actions taken by Russia to protest sanctions imposed by different countries across the globe. As a result, there has been an increase in food insecurity, and derailment in efforts to transition to cleaner energy. While political instability is a major setback to the achievement of GND, examples of bipartisan cooperation offer hope for progress. For instance, the United Kingdom's Climate Change Act, passed with cross-party support, sets legally binding targets for emissions reductions (UK Government, 2008). Similarly, New Zealand's Zero Carbon Act, endorsed by both major parties, establishes a framework for long-term climate policy (New Zealand Parliament, 2019). These examples showcase that it is possible for countries to set aside their differences for the common goal of both the environment and its citizenry. However, it is noteworthy that overcoming political polarization requires a multifaceted approach that emphasizes shared values and mutual interests Scott et al. (2020). This notion is emphasized at great length. For instance, in the United States, organizations like the Climate Leadership Council have been advocating for a bipartisan carbon pricing solution to address climate change (Climate Leadership Council, 2021), and this could be potent in overcoming cross-border challenges in the implementation of GND. Additionally, grassroots movements and community-based initiatives can mobilize support for climate action at the local level, transcending partisan divisions, especially in proposing options that would lead to customizable solutions not only to climate change, but to social inequalities and injustices (UNFCCC, 2019). By fostering dialogue, promoting cooperation, and building coalitions across ideological lines, policymakers can find common ground on the GND and advance solutions to the climate crisis.

4 Way Forward for Green Development

The Green New Deal (GND) movements across the world have ignited a global conversation about the feasibility and urgency of transitioning towards a sustainable future. In fact, such conversations seem to have sparked some increased vigour even among less developed economies to voice their concerns and demand a fair share in the climate action debate. While this bold vision carries the potential to address pressing environmental, economic, and social challenges, it simultaneously faces a whirlwind of debates and controversies, more so in regard to financing and multilateral bodies recalibration. Navigating this complex landscape and guaranteeing continued momentum requires a closer examination of the key questions, transformative potential, and practical considerations associated with Green New Deals.

Among the key questions on the road ahead might include:

1. Is there a need for a collective imperative for international cooperation? The answer to this question is debatable, but to set the tone, it is worth noting that achieving ambitious climate goals on a global scale necessitates an unprecedented level of international cooperation (UN Climate Change, 2022). Looking at the history of proposed climate action; more so the gains and misses already witnessed, it is arguable that there is more potential in unity than in division. For instance, in the Paris Agreement, it was proposed that by 2020, developed countries could contribute resources to raise at least $100 billion to support less developed and vulnerable economies. While this was not realized, there was a substantial effort (with approximately $80 billion raised), that significantly contributed to building resilience and adaptability in different economies. However, it must be appreciated that more needs to be done. Regarding the GND,

while successful examples like the European Green Deal (Fit for 55) (European Council, 2021) showcase the potential of regional cooperation, achieving truly transformative change likely requires more comprehensive global collaboration. This raises questions about how to overcome historical disparities in national commitments, forge effective and just international institutions, and ensure equitable burden-sharing across developed and developing nations.

While ambitious international collaboration presents undeniable potential for addressing global climate challenges, the complexity of aligning various national interests and development stages cannot be understated. True international cooperation extends beyond financial contributions and involves a holistic integration of technological, intellectual, and infrastructural resources. This integration must cater not only to immediate climate resilience but also to the sustainable economic development of less developed nations. In this context, the role of technology transfer and capacity-building becomes crucial. Developed nations must commit to not only sharing resources but also technology and expertise that can empower developing nations to leapfrog to cleaner, more sustainable technologies without repeating the environmental degradation historically associated with industrialization. Furthermore, international agreements should include mechanisms for monitoring and enforcing compliance, ensuring that commitments are not only made but also faithfully executed. Moreover, the global cooperation framework must consider the socio-economic impacts of climate policies. Policies designed in affluent nations might not be suitable for poorer regions where basic needs are still a pressing concern. Therefore, a flexible, context-sensitive approach that respects local conditions and priorities is vital for the success of any global strategy. This may involve creating customized pathways for different countries, acknowledging that while the ultimate goal of climate resilience is common, the journey there might look different for each nation.

2. Is just transition feasible, or a mirage? From the discussion above, it has been elaborated that a just transition to a green economy demands robust strategies that protect vulnerable communities and workers potentially impacted by the shift. This requires addressing concerns about job losses in fossil fuel-dependent industries through targeted reskilling programs, social safety nets, and investments in new green sectors. Additionally, ensuring equitable access to the

benefits of green jobs and a sustainable economy with the capacity to empower marginalized communities and address historical inequalities, including in trade of balance and their share of financial support targeting climate action (Walker, 2022). While the concept of a just transition is compelling and morally imperative, its implementation is fraught with complexities.

One of the major challenges lies in the balance between environmental imperatives and economic realities. For many regions dependent on fossil fuels, the transition not only threatens jobs but also local economies. To navigate this, transition strategies must be both region-specific and inclusive, involving stakeholders from affected communities in the planning and decision-making process to tailor solutions that are both practical and acceptable to those impacted. Moreover, the just transition framework must be integrated into broader economic policies. Governments should consider how environmental policies intersect with economic and social policies to avoid unintended consequences that could exacerbate inequalities. This might include revising tax policies, adjusting welfare systems, and even reshaping education and training programs to prepare the workforce for new industries. Additionally, international support plays a critical role, especially for countries lacking the financial resources to implement such comprehensive transitions. Developed countries and international financial institutions can support these efforts through funding, partnerships, and shared expertise. Furthermore, the private sector's involvement through responsible investment and innovation in green technology is crucial to creating economic opportunities that can replace those lost in the transition.

3. Is it possible to bridge the divide by ensuring effective and successful engagement across the ideological spectrum? From the discussion above, it is evident that the political landscape surrounding GNDs is often characterized by deep polarization and self-centredness. Overcoming this then requires innovative communication strategies that move beyond partisan rhetoric and focus on shared values like clean air, healthy communities, and economic opportunities. It calls for building coalitions across ideological divides by adopting strategies like framing proposals in terms of national security and energy independence and appealing to both

progressive and moderate parties. It also entails fostering constructive dialogue and collaborative decision-making processes aimed at exposing controversies and proposing potent solutions, as well as building broader public support.

Efforts to engage various ideological groups effectively also demand an emphasis on education and outreach. Public understanding of what GNDs entail and how they can benefit society at large—regardless of political or ideological affiliation—is crucial. By demystifying the goals and mechanisms of GNDs through accessible and transparent communication, stakeholders can reduce misinformation and scepticism. Furthermore, engagement strategies should include robust feedback mechanisms that allow communities to voice concerns and contribute ideas. This participatory approach not only enhances the legitimacy of GND policies but also increases the likelihood of their acceptance and success. Engaging local leaders and influencers from various sectors—religious, cultural, business, and educational—can amplify this effect, helping to translate the high-level policy language of GNDs into concrete, relatable actions that resonate with the everyday experiences of individuals. In addition, fostering bipartisanship through shared economic interests, such as job creation in new green industries, can provide a common ground for dialogue. Economic incentives such as tax benefits, grants, and subsidies for green projects can attract a wider range of political support, aligning stakeholders with diverse ideologies towards a common goal of sustainable economic development.

4. Will it ever be possible to find the right balance in sustainable financing? This question is premised on the fact that the significant investments required for GNDs raise concerns about funding models and their potential fiscal implications (Walker, 2022). For instance, in the recent past, countries, especially in the global south have started to demand a fair share in financial support, especially in view of the expansive green resources they possess and their inherent potential to lead the way in climate change action (Colenbrander et al., 2022). Innovative strategies like public–private partnerships, carbon pricing mechanisms, and green bonds can play a crucial role in mobilizing necessary capital (Fischer & Jacobsen, 2021). Moreover, restructuring the existing international financial instruments and leveraging private sector investments can further expand the available resources. However, striking a balance between ambitious goals,

fiscal responsibility, and individual burden remains a key challenge requiring careful consideration and transparent communication and collaboration.

Finding the right balance in sustainable financing also involves enhancing the efficiency and accountability of how funds are allocated and used. Ensuring that green financing is not only accessible but also yields measurable environmental and economic benefits is essential. This could be supported by adopting more stringent monitoring and evaluation frameworks that track the impact of funded projects, thereby increasing transparency and building trust among stakeholders.

To address the challenge of equity in financing, especially for under-resourced regions, there should be more tailored financial instruments. These instruments could include risk guarantees and concessional loans that lower the barrier to entry for green investments in developing countries. Additionally, a global green fund could be established, specifically designed to support those countries with rich natural resources but limited financial capacity to exploit these resources sustainably. Furthermore, the integration of sustainability criteria into the broader financial system could encourage more investments in green projects. This might involve adjusting financial regulations to favour investments in sustainability, such as lower reserve requirements for green bonds or preferential interest rates for loans financing renewable energy projects. Moreover, enhancing international cooperation in financial regulation can prevent the offshoring of environmentally harmful practices to less regulated markets. This requires a concerted effort from the global north and south to harmonize standards and share best practices in green financing, ensuring that global financial flows support a broad transition to a sustainable economy.

5. Will it be possible to measure progress to facilitate accountability and improvement? Effective tracking of the progress and impact of GNDs necessitates robust monitoring and evaluation frameworks (Calhoun & Fong, 2022). These frameworks should encompass quantitative metrics like emissions reductions, green job creation, and economic growth, while also incorporating qualitative assessments of social justice, community well-being, and ecosystem health. Regular transparent reporting and ongoing evaluation are and will continue to be crucial for holding accountable those responsible

for implementation and ensuring continuous improvement based on data-driven insights. In addition to established frameworks, the development of new technologies such as blockchain could significantly enhance the transparency and traceability of green initiatives. By utilizing decentralized and tamper-proof digital ledgers, stakeholders can more reliably track the flow of funds and the resultant environmental impacts, minimizing the risk of corruption and mismanagement. Further, the integration of big data analytics and artificial intelligence can refine the precision of monitoring and evaluation. These technologies can process vast amounts of data from multiple sources to provide more nuanced insights into the effectiveness of policies and projects. For example, satellite imagery and sensor data can be analysed to track deforestation rates, carbon sequestration, or the proliferation of renewable energy installations in real time.

Community involvement in the monitoring process also adds a crucial layer of oversight and engagement. Local citizen groups equipped with the right tools can collect data on local environmental conditions, providing a grassroots perspective that may be overlooked in higher-level analyses. This participatory monitoring approach not only empowers communities but also enhances the granularity and accuracy of the data collected. Additionally, establishing international standards for reporting and accountability can help ensure consistency and comparability between regions and countries. This could involve creating a set of universal indicators that all parties to international climate agreements would be required to measure and report publicly.

While the above questions remain potent, it should never be lost that the GND offers transformative potential across various dimensions. For instance, all the proposed ones by different countries offer vivid decarbonization pathways, particularly regarding transitioning to renewable energy sources and implementing energy efficiency measures. GND are also hailed for their potential to promote the massive creation of green Jobs. From the discussion above, it has been elaborated that investments in clean energy, sustainable infrastructure, and resource conservation can create millions of new jobs in diverse sectors, contributing to economic growth and tackling unemployment (Scott et al., 2020). These green jobs often offer competitive wages, improved working conditions, and opportunities for career advancement, leading to a more prosperous and equitable workforce.

GNDs are also seen to have significant potential in the enhancement of public health through reducing air and water pollution, greening urban spaces, and protecting biodiversity (Green, 2022). This translates to tangible benefits like lower rates of respiratory illnesses, cardiovascular diseases, and other environmental health risks, contributing to healthier communities and increased life expectancy. Further, GNDs have significant potential in resilience building, particularly in view of their emphasis on investing in green infrastructure, natural capital conservation, and early warning systems that can help communities adapt to the inevitable impacts of climate change, such as extreme weather events and rising sea levels. This proactive approach enhances resilience, protects livelihoods, and minimizes economic losses associated with climate disasters.

References

Allam, Z. (2020a). *The Rise of Autonomous Smart Cities: Technology, Economic Performance and Climate Resilience*. Springer Nature.

Allam, Z. (2020b). *Surveying the Covid-19 Pandemic and Its Implications: Urban Health*. Elsevier Science.

Allam, Z., & Jones, D. (2019a). Climate Change and Economic Resilience through Urban and Cultural Heritage: The Case of Emerging Small Island Developing States Economies. *Economies*, 7.

Allam, Z., Jones, D., Biyik, C., et al. (2021a). Rewriting City Narratives and Spirit: Post-pandemic Urban Recovery Mechanisms in the Shadow of the Global 'Black Lives Matter' Movement. *Research in Globalization, 3*, 100064.

Allam, Z., Jones, D., & Thondoo, M. (2020a). *Cities and Climate Change: Climate Policy, Economic Resilience and Urban Sustainability*. Springer Nature.

Allam, Z., Jones, D., & Thondoo, M. (2020b). *Cities and Climate Change: Climate Policy*. Springer International Publishing.

Allam, Z., Jones, D., & Thondoo, M. (2020c). Climate Change Mitigation and Urban Liveability. In Z. Allam, D. Jones, & M. Thondoo (Eds.), *Cities and Climate Change: Climate Policy, Economic Resilience and Urban Sustainability* (pp. 55–81). Springer International Publishing.

Allam, Z., Jones, D., & Thondoo, M. (2020d). Decarbonization and Urban Sustainability. *Cities and Climate Change*, 33–54.

Allam, Z., & Jones, D. S. (2019b). Attracting Investment by Introducing the City as a Special Economic Zone: A Perspective From Mauritius. *Urban Research & Practice, 12*, 201–207.

Allam, Z., & Jones, D. S. (2021). Future (post-COVID) Digital, Smart and Sustainable Cities in the Wake of 6G: Digital Twins, Immersive Realities and New Urban Economies. *Land Use Policy, 101*, 105201.

Allam, Z., Jones, D. S., Roös, P. B., et al. (2022a). "Quarantined Within a Quarantine": COVID-19 and GIS Dynamic Scenario Modeling in Tasmania, Australia. In *Data Science for COVID-19* (pp. 355–395). Academic Press.

Allam, Z., Moreno, C., Chabaud, D., et al. (2022b). Proximity-Based Planning and the "15-Minute City": A Sustainable Model for the City of the Future. In *The Palgrave Handbook of Global Sustainability*.

Allam, Z., Sharifi, A., Giurco, D., et al. (2021b). On the Theoretical Conceptualisations, Knowledge Structures and Trends of Green New Deals. *Sustainability, 13*, 12529.

Allam, Z., Sharifi, A., Giurco, D., et al. (2022c). Green New Deals Could Be the Answer to COP26's Deep Decarbonisation Needs. *Sustainable Horizons, 1*, 100006.

Allam, Z., Cheshmehzangi, A., & Jones, D. S. (2023). *Climate and Social Justice: The Political Economy of Urban Resilience and Mercantilism*. Springer.

Atkin, E. (2019). *The Democrats Stole the Green Party's Best Idea*. https://newrepublic.com/article/153127/democrats-stole-green-partys-best-idea

Bloomfield, J., & Steward, F. (2020). The Politics of the Green New Deal. *The Political Quarterly, 91*, 770–779.

Bradshaw, C. J. A., Ehrlich, P. R., Beattie, A., et al. (2021). Underestimating the Challenges of Avoiding a Ghastly Future. *Frontiers in Conservation Science, 1*, 615419.

Calhoun, C., & Fong, B. Y. (2022). Introduction. In C. Calhoun & B. Y. Fong (Eds.), *The Green New Deal and the Future of Work* (pp. 1–20). Columbia University Press.

Carbon Brief. (2023). *Analysis: Global CO_2 Emissions Could Peak as Soon as 2023, IEA Data Reveals*. https://www.carbonbrief.org/analysis-global-co2-emissions-could-peak-as-soon-as-2023-iea-data-reveals/#:~:text=The%20IEA's%20latest%20World%20Energy,will%20happen%20%E2%80%9Cby%202025%E2%80%9D

Ceola, S., Mård, J., & Di Baldassarre, G. (2023). Drought and Human Mobility in Africa. *Earth's Future, 11*, e2023EF003510.

Chabaud, D., Allam, Z., Gall, C., et al. (2023). Redefining Investable Infrastructure in Developing Nations in a Postpandemic Era: The Case of the 15-Minute City. In *Resilient and Sustainable Cities* (pp. 136–149). Elsevier.

Chatzky, A., & Siripurapu, A. (2021). *Envisioning a Green New Deal: A Global Comparison*. https://www.cfr.org/backgrounder/envisioning-green-new-deal-global-comparison

Cheshmehzangi, A. (2020). The Analysis of Global Warming Patterns from 1970s to 2010s. *Atmospheric and Climate Sciences, 10*(3). https://doi.org/10.4236/acs.2020.103022

Cheshmehzangi, A., & Chen, H. (2021). *China's Sustainability Transitions: Low Carbon and Climate-Resilient plan for Carbon Neutral 2060*. Springer.

Cheshmehzangi, A., & Dawodu, A. (2019). *Sustainable Urban Development in the Age of Climate Change—People: The Cure or Curse*. Palgrave Macmillan.

Cheshmehzangi, A., You, N., Siri, J., & Birch, E. (2024). *Harnessing Urban Innovation to Unlock the Sustainable Development Goals*. Springer.

Chiang, F., Mazdiyasni, O., & AghaKouchak, A. (2021). Evidence of Anthropogenic Impacts on Global Drought Frequency, Duration, and Intensity. *Nature Communications, 12*, 2754.

Chohan, U. (2019). A Green New Deal: Discursive Review and Appraisal. *SSRN Electronic Journal*.

Chowdhury, S. (2021). *South Korea's Green New Dealinthe Year of Transition*. https://www.undp.org/blogs/south-koreas-green-new-deal-year-transition

Climate Leadership Council. (2021). *Our Carbon Dividends Plan*. https://clcouncil.org/four-pillars/#:~:text=This%20plan%20will%20position%20all,energy%20future%20belongs%20to%20everyone

Colenbrander, S., Pettinotti, L., & Cao, Y. (2022). *A Fair Share of Climate Finance? An Appraisal of Past Performance, Future Pledges and Prospective Contributors* (ODI Working Paper, pp. 1–64).

Congress. (2019). 116th Congress 1st Session: H. RES. 109. In: Representative Ho (ed). Congress.

Conte, N. (2019). *Green New Deal*. https://wtpartnership.co/green-new-deal/

Dartford, K. (2021). *How Can the European Green Deal Avoid 'Potential Frictions' With Other Countries?* https://www.euronews.com/green/2021/02/03/how-can-the-european-green-deal-avoid-potential-frictions-with-other-countries

Davies, R. (2021). *Western Europe—Storms Case Floods in Switzerland, Germany and UK*. https://floodlist.com/europe/storms-floods-switzerland-germany-uk-july-2021

DiEM25. (2019). *The Green New Deal for Europe*. https://diem25.org/campaign/green-new-deal/

Dofonsou, S. G., Chikoye, D., Alene, A., et al. (2023). Strategic Foresight Analysis on Droughts in Southern Africa and Its Implications for Food Security. *Frontier Sustainable Food System*, 7.

Donato, D. (2020). *The Green New Deal and China's Green Revolution*. https://www.peoplesworld.org/article/the-green-new-deal-and-chinas-green-revolution/

Doreen, E. M. (2014). The Right to Be Free of Fear: Indigeneity and the United Nations. *Wicazo Sa Review, 29*, 63–87.

European Commission. (2019). *The European Green Deal: Questions & Answers*. https://ec.europa.eu/commission/presscorner/detail/en/QANDA_19_6690

European Council. (2021). *Fit for 55*. https://www.consilium.europa.eu/en/policies/green-deal/fit-for-55-the-eu-plan-for-a-green-transition/#:~:text=Fit%20for%2055%20refers%20to,line%20with%20the%202030%20goal

Feitelson, E., Tamimi, A., & Rosenthal, G. (2012). Climate Change and Security in the Israeli-Palestinian Context. *Journal of Peace Research, 49*, 241–257.

Fischer, C., & Jacobsen, G. D. (2021). The Green New Deal and the Future of Carbon Pricing. *Journal of Policy Analysis and Management, 40*, 988–995.

French, H., Renner, M., & Gardner, G. (2009). *Toward a Transatlantic Green New Deal: Tackling the Climate and Economic Crises*. Heinrich-Böll-Stiftung.

Google Trends. (2021). *Green New Deal*. https://trends.google.com/trends/explore?date=all_2008&gprop=images&q=green%20new%20deal

Google Trends. (2023). *New Green Deal*. https://trends.google.com/trends/explore?date=2008-01-01%202023-12-31&q=new%20green%20deal

Government of New Zealand. (2022). *Aotearoa New Zealand Emissions Reduction Plan* (pp. 1–348). Ministry for the Environment.

Government of Singapore. (2021). *Introducing the Green Plan*. Available at: Green New Deal Singapore.

Government of Spain (2020). *Strategic Energy and Climate Framework* (pp. 1–62). Instituto para la Transicion Justa, O. A.

Green, J. (2022). Greening Keynes? Productivist Lineages of the Green New Deal. *The Anthropocene Review, 9*, 324–343.

Green Party. (2021) *The Green New Deal*. https://www.gp.org/green_new_deal

Green Party of the United States. (2006). *The Green New Deal*. https://gpus.org/organizing-tools/the-green-new-deal/

Gunia, A. (2021). *Could the World's Largest Green Energy Hub Redefine How Companies Work With Indigenous People?* https://time.com/6080182/australia-green-energy-aboriginal-people/

IEA. (2023). *CO_2 Emissions in 2022*. IEA.

Inman, E. N., Hobbs, R. J., & Tsvuura, Z. (2020). No Safety Net in the Face of Climate Change: The Case of Pastoralists in Kunene Region, Namibia. *PLoS ONE, 15*, e0238982.

Intergovernmental Panel on Climate Change. (2018). *Global Warming of 1.5°C: An IPCC Special Report on the Impacts of Global Warming of 1.5°C Above Pre-industrial Levels and Related Global Greenhouse Gas Emission Pathways, in the Context of Strengthening the Global Response to the Threat of Climate Change, Sustainable Development, and Efforts to Eradicate Poverty*. Intergovernmental Panel on Climate Change.

International Renewable Energy Agency. (2020). *Renewable Energy and Jobs—Annual Review 2020*. IRENA.

IPCC. (2021). The Physical Science Basis. In V. Masson-Delmotte, P. Zhai, A. Pirani, et al. (Eds.), *Contribution of Working Group I to the Sixth Assessment Report of the Intergovernmental Panel on Climate Change*. Cambridge University Press.

IPCC. (2023). Sections. In Core Writing Team, H. Lee, & J. Romero (Eds.), *Climate Change 2023: Synthesis Report. Contribution of Working Groups I, II and III to the Sixth Assessment Report of the Intergovernmental Panel on Climate Change* (pp. 35–115). IPCC.

Assessment Report of the Intergovernmental Panel on Climate Change. Geneva, Switzerland: IPCC, 35–115**.

Kozul-Wright, R. (2023). *A Climate Finance Goal That Works for Developing Countries*. https://unctad.org/news/climate-finance-goal-works-developing-countries

Le Quéré, C., Peters, G. P., Friedlingstein, P., et al. (2021). Fossil CO_2 Emissions in the Post-COVID-19 Era. *Nature Climate Change, 11*, 197–199.

League of Conservation Voters (LCV). (2019). *LCV Supports Green New Deal Resolution*. https://www.lcv.org/article/lcv-supports-green-new-deal-resolution/

London Assembly. (2021). *A Green New Deal*. https://www.london.gov.uk/coronavirus/londons-recovery-coronavirus-crisis/recovery-context/green-new-deal

MacArthur, J. L., Hoicka, C. E., Castleden, H., et al. (2020). Canada's Green New Deal: Forging the Socio-political Foundations of Climate Resilient Infrastructure? *Energy Research & Social Science, 65*, 101442.

Mastini, R., Kallis, G., & Hickel, J. (2021). A Green New Deal Without Growth? *Ecological Economics, 179*, 106832.

Ministry of Environment. (2020). *Green New Deal to Drive Green Recovery in Asia-Pacific Region*. http://eng.me.go.kr/eng/web/board/read.do?pagerOffset=0&maxPageItems=10&maxIndexPages=10&searchKey=&searchValue=&menuId=461&orgCd=&boardId=1415460&boardMasterId=522&boardCategoryId=&decorator=

Moreno-Brid, J. C., & Gallagher, K. (2020). Mexico's Road to a Green New Deal. *NACLA Report on the Americas, 52*, 152–157.

New Economics Foundation. (2008). *Green New Deal: Cutting Carbon Emissions, Boosting Nature and Creating Good Jobs*. https://neweconomics.org/about/our-missions/green-new-deal

New Zealand Parliament. (2019). *Climate Chage Response (Zero Carbon) Amendment Act 2019*. https://environment.govt.nz/acts-and-regulations/acts/climate-change-response-amendment-act-2019/

Paszak, P. (2020). *The Chinese Vision of the Green Deal*. https://ec.europa.eu/clima/sites/default/files/international/cooperation/china/docs/zhong_ou_he_zuo_yan_jiu_bao_gao_-20201128_en.pdf

Rawtani, D., Gupta, G., Khatri, N., et al. (2022). Environmental Damages Due to War in Ukraine: A Perspective. *Science of the Total Environment, 850*, 157932.

Robinson, E. (2018). *How Are We Doing on a Green New Deal?* https://www.opendemocracy.net/en/transformation/how-are-we-ng-on-green-new-deal/

Roy, P., Pal, S. C., Chakrabortty, R., et al. (2023). Effects of Climate Change and Sea-Level Rise on Coastal Habitat: Vulnerability Assessment, Adaptation Strategies and Policy Recommendations. *Journal of Environmental Management, 330*, 117187.

Saha, S, Subramanian, N., & Menzelos, A. (2020). *Memo: A Green Marshall Plan—America's Global Climate Compact*. https://www.dataforprogress.org/memos/green-marshall-plan

Scott, R. E., Mokhiber, Z., & Perez, D. (2020). *Rebuilding American Manufacturing—Potential Job Gains by State and Industry*. https://www.epi.org/publication/rebuilding-american-manufacturing-potential-job-gains-by-state-and-industry-analysis-of-trade-infrastructure-and-clean-energy-energy-efficiency-proposals/

Shan, Y., Ou, J., Wang, D., et al. (2021). Impacts of COVID-19 and Fiscal Stimuli on Global Emissions and the Paris Agreement. *Nature Climate Change, 11*, 200–206.

Táíwò, O. (2019). *How a Green New Deal Could Exploit Developing Countries*. https://theconversation.com/how-a-green-new-deal-could-exploit-developing-countries-111726

The Climate Mobilization. (2017). *About*. https://www.theclimatemobilization.org/about/

The European Commisision. (2020). *Circular Economy Action Plan: For a Cleaner and More Competitive Europe*. European Union.

The Greens/EFA. (2011). *The Green New Deal: More than a Policy, a Programme for Living*. https://www.greens-efa.eu/en/article/document/the-green-new-deal

The World Bank. (2020). *Solar Photovoltaic Power Potential by Country*. https://www.worldbank.org/en/topic/energy/publication/solar-photovoltaic-power-potential-by-country

UK Government. (2008). Climate Change Act 2008. In: Legislation (ed).

UN. (2021). *Green Economy*. https://sustainabledevelopment.un.org/index.php?menu=1446

UNEP. (2009). *Global Green New Deal: An Update for the G20 Pittsburgh Summit*. https://wedocs.unep.org/rest/bitstreams/11748/retrieve#:~:text=In%20response%20to%20the%20financial,change%2C%20environmental%20degradation%20and%20poverty

UNEP. (2014). *Emerging Issues for Small Island Developing States*. UNEP.

UNEP. (2021). *Are We on Track for a Green Recovery? Not Yet*. https://www.unep.org/news-and-stories/press-release/are-we-track-green-recovery-not-yet

UNFCCC. (2019). *Cities and Local Action to Combat Climate Change*. https://unfccc.int/topics/education-youth/youth/global-youth-video-competition/global-youth-video-competition-2019/cities-and-local-action-to-combat-climate-change

UNFCCC. (2021). Nationally Determined Contributions under the Paris Agreement: Synthesis Report by the Secretariat. *Conference of the Parties Serving as the Meeting of the Parties to the Paris Agreement*. United Nations Framework Convention on Climate Change.

United Nations. (2022). A Just Green Transition: Concepts and Practice So Far. In *Future of the World: Policy Brief* (pp. 1–6). United Nations Department of Economic and Social Affairs.

United Nations Development Programme (UNDP). (2015). *Sustainable Development Goals*. UNDP.

United Nations Framework Convention on Climate Change. (2015). Synthesis Report on the Aggregate Effect of the Intended Nationally Determined Contributions. In *Report of the Ad Hoc Working Group on the Durban Platform for Enhanced Action* (p. 66). UNFCCC.

Visentin, L. (2020). *'Massive Green Deal': NSW Environment Minister Spruiks $2 Billion Energy Package*. https://www.smh.com.au/politics/nsw/massive-green-deal-nsw-environment-minister-spruiks-2-billion-energy-package-20200131-p53wgg.html

Walker, R. A. (2022). From the New Deal to the Green New Deal. In C. Calhoun & B. Y. Fong (Eds.), *The Green New Deal and the Future of Work* (pp. 23–52). Columbia University Press.

Zaheer A and Society. (2020). Sustainability and Resilience in Megacities through Energy Diversification, Land Fragmentation and Fiscal Mechanisms. *Sustainable Cities, 53*, 101841.

Zhou, J., Yuan, Y., Fu, Z., et al. (2022). The Impact of Public Health Events on Green Economy Efficiency in the Context of Environmental Regulation. *Frontiers in Public Health, 10*, 996139.

CHAPTER 2

Theoretical and International Perspectives on Sustainable Futures

Abstract This chapter explores theoretical and international perspectives on sustainable futures, focusing on ecological economics, social ecology, and environmental justice. Sustainable development is analysed through environmental preservation, social justice, and economic viability lenses. Ecological economics challenges traditional paradigms by emphasizing the finite nature of planetary resources and prioritizing long-term resilience over short-term profit. Social ecology examines the links between societal structures and environmental degradation, advocating decentralized and democratic means to harmonize human communities with the natural world. Environmental justice addresses the disproportionate environmental harm experienced by marginalized communities, promoting equitable distribution of environmental risks and benefits. The Green New Deal (GND) is presented as a policy framework that integrates these theoretical approaches to address climate change and economic inequality. By incorporating the principles of ecological economics, social ecology, and environmental justice, the GND offers a holistic approach to sustainable development. This chapter highlights the necessity of international cooperation and the exchange of best practices to achieve global sustainability goals. The synthesis of these frameworks reveals that sustainable futures depend on reimagining economic development, fostering social inclusivity, and preserving environmental integrity. The chapter emphasizes the importance of innovative governance structures, equitable resource distribution, and a redefined relationship between humanity and

Z. Allam and A. Cheshmehzangi, *Sustainable Futures and Green New Deals*, https://doi.org/10.1007/978-3-031-63642-4_2

the natural world. By moving beyond siloed approaches and embracing interconnected strategies, pathways towards sustainable futures can be effectively forged.

Keywords Sustainable futures · Ecological economics · Social ecology · Environmental justice · Green New Deal (GND) · Climate change · International cooperation · Economic viability

1 Introduction to Sustainable Futures

The concept of sustainable futures demands a nuanced understanding, particularly in the face of increasing global challenges such as climate change, biodiversity loss, and socio-economic disparities and vulnerability of economies such as SIDS (Allam & Jones, 2019; Zaheer & Jones, 2018). This discourse necessitates an interrogation of sustainable development through the lens of not only environmental preservation but also through the prisms of social justice and economic viability. The theoretical scaffolding of ecological economics, social ecology, and environmental justice discussed in this chapter provides a robust framework for dissecting the complexities of sustainability. These paradigms collectively underscore the imperative of an integrated approach, intertwining ecological integrity with socio-economic equity, to forge pathways towards sustainable futures.

Ecological economics, as a transdisciplinary field, departs from conventional economic thought by positing the economy as a subsystem of the environment rather than its master. This perspective foregrounds the finite nature of planetary resources, advocating for a paradigm shift towards valuing ecological stability and long-term resilience over short-term profit maximization. The framework of social ecology, pioneered by Bookchin (1982) extends this discourse by examining the interlinkages between hierarchical societal structures and environmental degradation, proposing a reharmonization of human communities with the natural world through decentralized and democratic means.

The conversation on social ecology is further enriched by the concept of environmental justice emerging from grassroots movements against environmental racism and inequity (Allam et al., 2023). Environmental justice emphasizes the disproportionate burden of environmental harm borne by marginalized communities. It also argues for equitable distribution of environmental risks and benefits (Bullard, 1990). The synthesis of these concepts provides a comprehensive lens through which the pursuit of sustainability can be envisioned as an inherently multi-dimensional pursuit aimed at integrating ecological health, social equity, and economic inclusivity, as proposed in the diverse Green Deals fronted in different countries and jurisdictions.

In summary, GNDs are policy propositions that have been gaining traction in several countries over the last decade. They are geared towards an integrated approach to sustainability, social justice, and economic equality (Allam, Sharifi, et al., 2022; Walker, 2022). By aiming to address climate change and economic inequality simultaneously (Cheshmehzangi, 2020; Cheshmehzangi et al., 2024), the GND represents a paradigmatic shift towards a holistic framework for sustainable development (Calhoun & Fong, 2022). Its emphasis on renewable energy transition, job creation in sustainable sectors, and social welfare enhancements encapsulates the essential elements of ecological economics, social ecology, and environmental justice (Cheshmehzangi & Dawodu, 2019; United Nations, 2022a). However, the attainment of a sustainable future transcends national borders and requires global collaborative efforts which are beyond partisan and selfish endeavours. This is true, noting that previous global collaborations on climate agendas such as the Paris Agreement have experienced unprecedented hurdles due to non-committal altitude by key players. From a wide range of literature, it has been made clear that the interdependent nature of ecological systems, coupled with the globalized structure of the economy, necessitates cross-border cooperation to address transnational environmental issues and to facilitate the exchange of best practices and technologies. The imperative for non-partisan collaborations has also been explicitly made clear in major international frameworks such as the Paris Agreement, the Sustainable Development Goals (SDGs) the New Urban Agenda (NUA) (United Nations, 2017).

In synthesizing these theoretical frameworks and considering the GND's approach, it becomes evident that sustainable futures hinge on the human ability to reconceptualize economic development, pursue and intently foster social inclusivity, and preserve environmental integrity through collaborative and transdisciplinary efforts (Fig. 1). From the experience of COVID-19 and other global pandemics (Allam, 2020a, 2020b, 2020c, 2020d, 2020e), it has become vivid that the human race is stronger when united rather than when acting as individuals or divided.

Therefore, even in the approach towards addressing complex global challenges such as climate change, social injustices, and economic inequality, there is need for a departure from siloed standpoints, advocating instead for a cohesive strategy that embraces the interconnectedness of social, economic, and environmental dimensions. As such, the pursuit of sustainable futures is not merely an environmental imperative but a comprehensive socio-economic and political endeavour, calling for innovative governance structures, equitable resource distribution, and a reimagined relationship between humanity and the natural world, as is well stipulated in diverse versions of GNDs borne in different parts of the world.

2 Ecological Economics: A Path to Sustainability

Ecological economics emerges as a transdisciplinary field that challenges the conventional paradigms of neoclassical economics by prioritizing the symbiosis between ecological systems and economic activities (Wironen & Erickson, 2020). This branch of economics which is distinct in its approach, emphasizes the finite nature of Earth's resources and advocates for a sustainable scale of economic activity within these biophysical limits (Gómez et al., 2023). Central to ecological economics foundational principles is the recognition of the economy as a subsystem of the larger, finite ecosystem, which fundamentally opposes the traditional economic view of endless growth and resource exploitation.

At its core, ecological economics hinges on several key principles: the acknowledgment of resource limits, the intrinsic value of ecosystem services, and the imperative of a just distribution of resources and opportunities. Unlike traditional economic models that externalize environmental degradation, ecological economics integrates the costs of resource depletion and ecosystem degradation into economic decision-making (Bartelmus, 1996). This integration is pivotal in recalibrating the economic systems towards sustainability especially in the current dispensation when the world is experiencing unprecedented challenges from increased emissions, more so as countries continue to struggle to regain their pre-COVID-19 economic growth momentums. The emissions are also being worsened by the heightened escalation of the war in Russia and Ukraine (Rawtani et al., 2022) and Israel and Palestine (United Nations, 2023a), and the hiccups in the supply chain of key renewable energy resources.

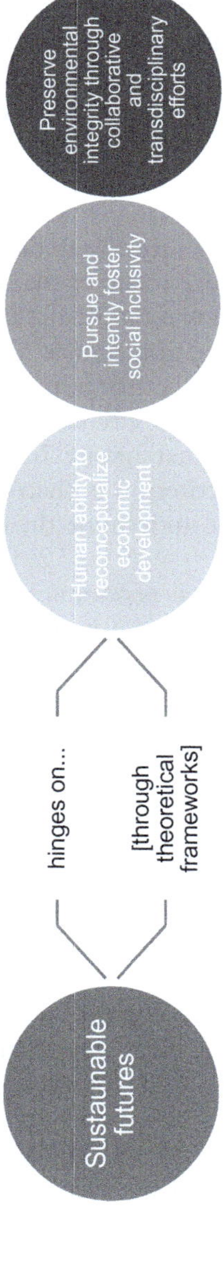

Fig. 1 Sustainable futures directions and basis of its theoretical frameworks

One of the most salient strengths of ecological economics is its focus on the valuation of ecosystem services. By attributing economic value to services provided by natural ecosystems, such as water purification, pollination, and climate regulation, ecological economics underscores the indispensable role of nature in sustaining human life and economic well-being. This valuation is instrumental in fostering policies and practices that prioritize ecosystem conservation and sustainable use of natural resources. Moreover, ecological economics champions the concept of 'strong sustainability', which stipulates that natural capital should be preserved intact owing to its non-substitutability. This contrasts with the 'weak sustainability,' concept which assumes that human-made capital can substitute natural capital. Even green infrastructures like renewable energy projects and sustainable transportation cannot be equated to the value of natural capital. Therefore, the best bet for humanity in promoting sustainability is to adopt concepts such as circular economy and build green infrastructures to complement what nature has already made available. In sum, a strong sustainability criterion is then crucial for ensuring that economic development does not undermine the ecological foundations it depends upon (Wilson, 2019).

Despite its strengths, ecological economics faces notable challenges. Implementation hurdles stem from the entrenched interests and structures of the existing economic system, which is predicated on continuous growth and resource exploitation. A case in point on entrenched interests is the ongoing Amazon deforestation crisis spurred by illegal enterprises (Boulton et al., 2022). Mainstream economic resistance to the fundamental shifts proposed by ecological economics is significant, as such shifts challenge conventional power structures and economic interests (Spash, 2008). Additionally, the operationalization of ecological economics principles, such as the valuation of ecosystem services, is fraught with methodological difficulties such as traditional pricing systems, which is not the most appropriate approach for unique and irreplaceable ecological assets. Assigning monetary value to complex and often intangible ecological functions can be challenging and contentious, potentially leading to an underestimation of the true value of these services (Brown et al., 2007).

The GND, with its emphasis on addressing climate change through transformative economic and social policies, aligns closely with the principles of ecological economics. The GND's focus on renewable energy transition, job creation in sustainable and green industries, and the reduction of greenhouse gas emissions mirrors ecological economics' advocacy for an economy that operates within Earth's biophysical limits while promoting social justice.

Ecological economics provides a theoretical foundation for the GND by emphasizing the need for an economic system that prioritizes environmental sustainability and equity (Neal, 2021). It advocates for a shift away from fossil fuels to more sustainable options such as renewable energies and alternative mobility options a move that is central to the GND. Further, it calls for the incorporation of external environmental costs into economic decision-making, a principle that is implicit in the different versions of GND's strategy for sustainable economic development.

Several case studies exemplify the successful application of ecological economics principles. For instance, the payment for ecosystem services (PES) programs in countries like Costa Rica have demonstrated how economic incentives can be effectively used to conserve biodiversity and sustain ecosystem services (Le et al., 2024). To put that in perspective, it is estimated that the $524 million invested in PES in Costa Rica has helped over 18,000 households as well as helped conserve approximately 1.3 million hectares of land (United Nations, 2023b). Another example is the transition town movement pioneered in Totnes, England which demonstrates how local economies can be restructured to focus on resilience, sustainability, and community well-being, reflecting ecological economics' emphasis on scale and localization (Barry & Quilley, 2009).

To inform global efforts toward sustainable futures, ecological economics advocates for several key strategies. These include the implementation of green taxes to internalize environmental externalities such as disincentivize environmental harmful activities and shifting behaviour towards sustainability. The taxes further help to generate revenue for environmental initiatives as well as foster the levelling of the playing ground for sustainable businesses by encouraging innovations in green technologies and emphasizing fair competition (Tan et al., 2022).

Another strategy is the adoption of degrowth strategies in overdeveloped economies to reduce environmental impact, and the promotion of circular economy models to minimize waste and maximize resource efficiency (Hoehn et al., 2021). A practical case that can be cited to demonstrate this is Germany's resource efficiency program that implements measures to decouple economic growth from ecological resource use, tackling waste, and promoting circularity (Government of Germany,

2020). Furthermore, ecological economics calls for strengthening international cooperation to address global environmental challenges, such as climate change and biodiversity loss, through mechanisms that ensure fair and equitable participation of all countries in sustainability efforts.

In summary, ecological economics stands as a beacon of hope for the Earth's ecological resources. While challenges to its implementation exist, its strengths—valuing nature, advocating 'strong sustainability', and aligning with transformative movements like the Green New Deal—offer a clear path towards a promising future, for both humanity and the ecology. Therefore, implementing green taxes, promoting degrowth and circularity, and fostering international cooperation can help harness the power of ecological economics to build a future where economy and ecology flourish in harmony.

3 Social Ecology: Communities and Sustainability

The perspective of social ecology was pioneered by Murray Bookchin in the 1960s and advocated for a comprehensive view of environmental problems as deeply entwined with social issues, arguing that ecological issues cannot be effectively addressed without confronting social injustices (Bookchin, 1982). It presents a compelling framework for understanding the intricate relationships between societal structures environmental issues, and social ecology. It also challenges the reductionist approaches that isolate environmental problems from their social contexts, instead promoting a holistic view that encompasses human communities, their social structures, and the natural world.

In relation to the modern world, the principal strength of social ecology lies in its holistic approach to human–environment interactions. That is, by integrating considerations of hierarchy, domination, and social inequality into environmental discourse, social ecology offers a nuanced understanding of how these social dimensions contribute to ecological degradation (Bookchin, 2007). This framework posits that the roots of environmental crises are not merely technical or economic but are fundamentally linked to the ways societies are organized and governed. A case in point to elaborate on this is the look at the Flint, Michigan water crisis that occurred after the water management body decided to switch its water source from Detroit's treated water to the Flint River. The motivation for this was to save money. However, since the river was not properly treated, it prompted lead to leach into peoples' homes, with the most affected being the Blacks and low-income residents, pointing to environmental injustice (Pauli, 2020).

Furthermore, social ecology emphasizes the potential for human agency in creating sustainable communities. It advocates for decentralized, democratic forms of governance and a reharmonization of human society with ecological principles through direct action and community engagement (Venturini et al., 2019). This emphasis on grassroots mobilization and the empowerment of local communities underscores the transformative potential of social ecology in addressing environmental challenges (Hummel et al., 2017).

Despite its positive influence, social ecology still faces some criticism. For instance, critics argue that its broad sweep could lead to an oversimplification of complex socio-economic issues. Others argue that the framework's focus on decentralization and communalism may not always adequately address the global scale of environmental problems or the intricacies of international economic systems. Additionally, the practical application of social ecology's ideals to diverse and complex modern societies remains a challenge, raising questions about the scalability and effectiveness of its proposed solutions. Among the potent challenges include funding of local projects, effective capacity-building and community engagement, and the ability to replicate successful projects in other areas noting their unique nature.

In the modern dispensation, social ecology significantly influences contemporary environmental movements, particularly in framing responses to climate change and environmental degradation. For instance, its emphasis and calls for a radical reorganization of society towards more egalitarian and ecologically harmonious structures resonates with the principles underlying the GNDs and other sustainability efforts (Calhoun & Fong, 2022). Social ecology's advocacy for community-led environmental initiatives and its critique of capitalist modes of production aligns with the Green New Deal's emphasis on social justice and economic restructuring for environmental sustainability.

Several community-based projects worldwide draw inspiration from social ecology, demonstrating its principles in action. For instance, the Transition Towns movement, initiated in Totnes, UK, embodies social ecology's ethos by fostering community resilience, localizing economies, and reducing ecological footprints through grassroots initiatives (Barry & Quilley, 2009). Similarly, urban gardening and permaculture projects (for instance, Seed Bombs for Biodiversity Restoration project in India [Saikia et al., 2023]) in cities around the globe reflect social ecology's influence by promoting sustainable food systems, community empowerment, and ecological stewardship.

Social ecology's emphasis on communalism and ecological interdependence has significant implications for fostering cross-border collaborations. For example, the Nile Basin Initiative entails collaboration of 11 African countries in the management of the river's resources, in conflict resolutions and sustainable water use practices. By highlighting the interconnectedness of environmental issues and social injustices across geographical boundaries, social ecology advocates for a cooperative approach to sustainability that transcends national borders (Bookchin, 2007). This perspective supports the development of international networks of solidarity and exchange, facilitating the sharing of knowledge, resources, and best practices among communities facing similar environmental challenges.

Ultimately, social ecology's ability to move beyond technical fixes and simplistic solutions is important in urging humanity to confront the deeper social roots of environmental problems, as it also focuses on socio-economic progress. It lays the groundwork for a future where the pursuit of environmental sustainability and social justice are not deemed as separate endeavours, but rather two interconnected sides. Embracing this holistic approach would offer a unique opportunity to forge a symbiotic relationship between humanity, biodiversity, and the Earth, ensuring a flourishing future for generations to come, just like is envisioned in initiatives such as the GND.

4 Environmental Justice: Equity in Sustainability

Environmental justice is a critical framework within the discourse on sustainability, underpinning the nexus between social equity and environmental health. It advocates for the just treatment and meaningful involvement of all people regardless of their demographics such as race, colour, nationality, or income status with respect to the development, implementation, and enforcement of environmental laws, regulations, and policies (Van Horne et al., 2023). In the sustainability discourse, its unique significance can be understood in how it focuses on addressing the disproportionate impact of environmental degradation and pollution on marginalized communities and vulnerable economies such as the Small

Island Developing States (SIDS) that often bear the greatest burdens of such adversities without equitable access to decision-making processes (Portelli et al., 2020).

The environmental justice framework can illuminate the systemic inequities that lead to marginalized communities suffering disproportionately from environmental degradation. For instance, using this framework, it is possible to expose the disproportionate impacts on SIDS vulnerable due to their uniqueness and limited resources that are not enough for retreat or adaptation. By helping to highlight diverse disparities borne by the aforementioned groups and countries, the framework fosters a more holistic understanding of sustainability that encompasses social dimensions alongside environmental considerations. The ultimate benefit of such an approach is not only seeking to protect the environment but also ensuring that all individuals have equal access to a healthy environment, thereby integrating social welfare into the core of environmental policy (Van Horne et al. 2023). Moreover, environmental justice champions the principle of participatory democracy, advocating for the involvement of affected communities in environmental decision-making processes. This inclusivity strengthens the legitimacy and effectiveness of environmental policies by ensuring that they are informed by the diverse perspectives and needs of those most impacted (Agyeman & Evans, 2004).

The environmental justice movement, though powerful, grapples with a dilemma of balancing act: how to promote economic growth and development without harming the environment, more so for already marginalized communities. In some instances, efforts to advance economic development can conflict with the environmental justice goals, particularly when such compromises the integrity of the environment; thus, disproportionately affecting marginalized communities. This dilemma highlights the complexity of achieving sustainability in a manner that is both environmentally sound and socially equitable (Pulido, 2000). Additionally, the implementation of environmental justice principles can encounter resistance and pushback from political and economic interests that benefit from the status quo. Therefore, navigating these power dynamics can be a major challenge that can hinder the effective integration of environmental justice in policies and practices (Dolšak & Prakash, 2018).

The GND represents a significant step towards integrating environmental justice principles into broad-based policy initiatives (Boyle et al., 2021). By addressing climate change through a lens of social equity, the Green New Deal seeks to redress historical injustices and ensure that the transition to a sustainable economy benefits all communities, particularly those historically marginalized (Walker, 2022). This integration underscores a recognition of the interconnectedness of environmental and social justice issues and the necessity of a comprehensive approach to addressing them (Rifkin, 2019). Several case studies can be adapted to exemplify the application of environmental justice principles in policy and practice. For instance, the Environmental Justice Executive Order 12898, signed by President Bill Clinton in 1994, mandated federal agencies to consider environmental justice in their actions, marking a significant policy acknowledgement of the issue (EPA, 1998). Furthermore, community-based initiatives, such as the fight against the siting of hazardous waste facilities in Warren County, North Carolina, have showcased the power of grassroots activism in advancing environmental justice goals (Cole & Foster, 2001). Internationally, the concept of environmental justice has inspired movements and policies aimed at protecting vulnerable communities from environmental harm. The Niger Delta in Nigeria, for example, has seen significant environmental justice activism aimed at addressing the impacts of oil extraction on local communities (Obi, 2010).

In summary, it is worth noting that the principles of environmental justice hold considerable promise for enhancing international sustainability efforts. By foregrounding the need for equity in environmental policies, environmental justice can inform international agreements and cooperation mechanisms, ensuring that they not only address environmental degradation but also promote social equity. For instance, incorporating environmental justice into the framework of the Paris Agreement could enhance its effectiveness by ensuring that climate action is equitably distributed and benefits all communities, particularly those most vulnerable to climate change impacts. Its integration also into the Sustainable Development Goals could help in the quicker realization of some of the goals such as SDG 11 and 14 among others.

5 The Green New Deal: A Convergence of Theories

The GND represents a transformative policy framework aimed at addressing climate change and economic inequality through a comprehensive suite of measures that include transitioning to renewable energy, creating green jobs, and ensuring social justice. Originating in the United

States (Walker, 2022), but currently having spread to different parts of the world, the GND draws inspiration from the New Deal of the 1930s, but with a focus on sustainability and equity to confront the dual crises of environmental degradation and social inequality (Puaschunder, 2021). The GND is informed by principles of ecological economics discussed above, emphasizing the need to operate within Earth's biophysical limits and to recognize the value of ecosystem services. As currently constituted, the different versions of GNDs seem to challenge the traditional economic paradigm of unlimited growth by advocating for a sustainable scale of economic activity that respects natural resource constraints. However, critics argue that despite the positive economic benefits the GND can bring forth, its ambitious goals may face challenges in quantifying the economic value of ecosystem services and integrating this into mainstream economic decision-making (Zolyomi et al., 2023). However, such challenges can be overcome going forward by developing innovative valuation methods that capture the true worth of the ecosystem services beyond what traditional market metrics signify. Secondly, by incentivizing and integrating those values into economic decision-making through informed policies and market-based mechanisms.

On the social front, GND is greatly influenced by the social ecology theory, whose footprints are evident in its emphasis on decentralizing power and fostering community-led initiatives in energy production (e.g. the Green New Deal for Europe), food systems, and housing (e.g. the Green New Deal for Public Housing Act in the United States which seek to invest approximately $172 billion in a ten year period) (Congress, 2021). Borrowing from the social ecology principles, the GND calls for a harmonious relationship between society and nature, promoting an ecological society where human actions contribute to the health of the planet. Like in the case of ecological economics, the scale of the GND's proposals when based on the social ecology could encounter limitations in fully realizing the decentralized, community-based approach advocated by social ecology, given the existing centralized political and economic structures. Such issues could however be countered by adopting strategies such as multi-level governance and empowering regional and state bodies to come up with local-based solutions.

Regarding the environment and its health, environmental justice principles have become central to the GND, which aims to address the disproportionate impact of environmental degradation and climate change on marginalized communities. A look at different versions of GND, such as that of the United States, it is evident that they have made significant strides to ensure that the transition to a green economy is inclusive and equitable and provide support for those most affected by environmental injustices. However, it should be noted that the effectiveness of these provisions in practice remains a critical area for scrutiny, as achieving true equity in environmental policy is an ongoing challenge that requires a bipartisan approach.

On the international scale, the GND has sparked both criticism and support. Critics argue that its ambitious targets (some quarters project it could cost approximately $93 trillion in tens for the U.S. version [Natter, 2019]) may be impractical or could impose undue economic burdens, particularly on industries reliant on fossil fuels. Others raise concerns about the feasibility of rapidly transitioning to renewable energy without significant technological advances or financial investments.

Conversely, supporters highlight the urgent need for bold action on climate change and social inequality, pointing to the GND as a model for global cooperation on sustainability. The framework has inspired similar initiatives in other countries, such as the European Green Deal (European Council, 2021), which shares the GND's objectives of mitigating climate change and fostering economic growth through sustainable means.

Examples of initiatives similar to the GND can be found in various countries. For instance, Germany's Energiewende (energy transition) policy seeks to shift from fossil fuels to renewable energy, demonstrating a commitment to sustainability while navigating the challenges of energy security and economic competitiveness (Rechsteiner, 2021). In South Korea, the Green New Deal aims to create jobs and stimulate economic recovery through green technology and renewable energy investments (Chowdhury, 2021). These international examples underscore the potential for the principles of the GND to be adapted and implemented in diverse contexts. However, they also highlight the complexities involved in balancing environmental sustainability with economic and social objectives.

The debate over the feasibility and adaptability of the GND's principles for global sustainability efforts centres on several key issues. Proponents argue that the GND provides a necessary blueprint for action in the face of the climate crisis, offering a path towards a sustainable and just global economy (Calhoun & Fong, 2022; Fischer & Jacobsen, 2021). Critics, however, question the practicality of its implementation, citing concerns over economic viability, technological readiness, and political will. The adaptability of the GND's principles to different national contexts is critical for its global relevance. While the framework's core objectives are universally applicable, the strategies for achieving them must be tailored to local conditions, resources, and cultural factors. This flexibility is essential for fostering international collaboration and for the GND to serve as a catalyst for global sustainability efforts.

6 International Collaboration and Best Practices

In the recent decades, following a myriad of challenges such as the global financial crisis, and the 2019 COVID-19 pandemic outbreak, it has become obvious that the world is better united, and collaborative. International collaboration is even more profound when founded in the quest to navigate complex global terrains such as the sustainability challenges currently crimpling diverse facets of our world fabric (United Nations Development Programme [UNDP], 2015). The United Nations (UN) and the World Bank stand at the forefront of this global endeavour, catalysing cooperation through initiatives like the Sustainable Development Goals (SDGs), the Paris Agreements (United Nations Framework Convention on Climate Change, 2015), the New Urban Agenda (United Nations, 2017), and a myriad of climate resilience projects among others. These institutions serve as crucibles for international dialogue, shaping policies and marshalling the necessary technical and financial support to breathe life into sustainability ventures across diverse nations. The collaborative ethos they manifest in projects that span borders, affirming the collective capacity to mitigate environmental adversities and forge a promising future for humanity and biodiversity. The European Union's INTERREG program (European Commission, 2021) and the pan-African Great Green Wall (United Nations Environment Programme, 2023) initiative are emblematic of such collaborative successes, underscoring the synergy between communal effort and sustainable development.

In this global medley of sustainability efforts, countries have adopted unique lenses through which they interpret and implement the principles of ecological economics, social ecology, and environmental justice. Bhutan, for instance, has woven the fabric of ecological sustainability and social welfare into its developmental ethos, as reflected in its Gross National Happiness index (Ura et al., 2012). This harmonization of natural and social capital mirrors the essence that makes up ecological economics. Conversely, the Nordic model, with its robust social welfare systems, participatory community engagement, and unwavering commitment to environmental stewardship, serves as a living testament to the principles of social ecology and environmental justice (Midttun et al., 2011). These national paradigms underscore the rich diversity of approaches to achieving sustainability, each informed by distinct cultural, economic, and environmental contexts.

Despite the strides made in international cooperation for sustainability, there still exist some notable hurdles that need to be overcome for full benefits to be realized. Political strife, economic inequality, and the sacrosanct nature of national sovereignty often pose formidable challenges to collaborative endeavours. A case in point is the intention to withdraw from the Paris Agreement change agenda in 2017 and formal withdrawal in 2020 the United States citing sovereignty concerns among other issues (UNFCCC, 2020). The economic chasm between developed and developing nations frequently results in divergent sustainability priorities and capacities, while geopolitical tensions stymie international dialogue and cooperation. The global north–south divide further complicates this landscape, with discordant views on environmental responsibility and resource allocation exacerbating efforts toward unified action (United Nations, 2020).

Confronting these obstacles calls for a multifaceted strategy to invigorate global collaboration toward a sustainable future. A critical step in this direction is the fortification of international governance frameworks to enhance their efficacy in facilitating cross-border dialogue and cooperation. This may entail the recalibration of existing institutions to ensure they are more inclusive and reflective of the diverse interests and challenges of developing nations as has been the agitation by some quarters, especially from the global south (Sidiropoulos, 2022). Moreover, the establishment of equitable financial instruments, such as Green Bonds (The World Bank, 2023) and climate funds (Mahat et al., 2019), is imperative to democratize access to the resources requisite for sustainability projects across the globe, more so for the less developed and vulnerable economies. These mechanisms can serve as vital lifelines, enabling nations to transcend financial limitations and embark on sustainability ventures, and adaptation and resilience-building programs (M. Z. Allam, 2019; Allam & Jones, 2019).

Another approach that can be adopted to foster the attainment of set goals including in GNDs is the intentional bridging of the technological divide between developed and developing countries by levelling the playing field (Zhou & Peng, 2023). This could be achieved by approaches such as technology transfer and capacity-building initiatives, offering pathways for knowledge exchange and skill development that can empower nations to implement sustainable solutions customized to their unique circumstances. This approach not only facilitates the dissemination of cutting-edge technologies and best practices but also fosters a spirit of mutual learning and cooperation (Yu, 2023).

The fabric of international collaboration in sustainability is woven with threads of shared challenges and aspirations. It is a tapestry that reflects the collective endeavour to chart a course toward a future where ecological balance, social equity, and economic vitality are in harmony. This vision demands a concerted effort from all nations, transcending parochial interests and embracing a global ethos of cooperation and solidarity.

7 Conclusion: Towards a Sustainable Future

The discourse on sustainable futures is profoundly enriched by the contributions of ecological economics, social ecology, and environmental justice (Kish & Farley, 2021) and their potential impact on the GND. Each framework offers unique insights into the complexities of sustainability, revealing the multifaceted nature of the challenges at hand. Ecological economics, with its emphasis on recognizing the biophysical limits of our planet, urges a re-evaluation of economic systems to prioritize sustainability and the well-being of future generations. This could be through approaches such as natural capital accounting, investing in natural infrastructures such as reclaiming wetlands and forests, and emphasizing on adoption of circularity. It champions the idea that true economic progress cannot be decoupled from the health of our environmental systems, hence, the above examples become almost mandatory.

On the other hand, it has been comprehensively discussed how social ecology brings to the fore the indispensable role of societal structures in either exacerbating or mitigating environmental degradation. It posits that a sustainable future is unattainable without addressing the hierarchical and exploitative relationships that permeate human societies, leading to obvious social injustices and inequalities.

On its part, environmental justice expands the dialogue on environmental sustainability and health to include the moral imperatives of equity and inclusivity, highlighting the disproportionate burden of environmental harm borne by marginalized communities and advocating for their rightful place in decision-making processes.

As has been posited in the above sections, the integration of these perspectives with international practices is paramount for the realization of global sustainability. The challenges of environmental degradation, social injustice, and economic inequality are not confined by geopolitical boundaries; they are global phenomena that demand a coordinated and cohesive response (Allam et al., 2020a, 2020b). SDGs set forth by the United Nations exemplify an international framework aimed at addressing these interconnected challenges (United Nations, 2020). However, the realization of these goals necessitates a deep engagement with the theoretical insights provided by ecological economics, social ecology, and environmental justice, ensuring that policies are not only effective but equitable and just.

The pursuit of a sustainable future is laden with challenges but also brimming with opportunities (Allam, Bibri, Jones, et al., 2022; Allam, Moreno, Chabaud, et al., 2022), especially from the concerted effort shown by the political class across the world in the recent years (World Economic Forum, 2023). As such, the prevailing environmental crises, while daunting, offer a catalyst for innovation and transformation. The transition to renewable energy sources (Z. Allam, 2019; Dhunny et al., 2019), the pursuit of alternative mobility options, the adoption of sustainable agricultural practices, and the move towards circular economies are examples of how challenges can be transformed into opportunities for sustainable development (Strielkowski et al., 2021). These initiatives not only mitigate environmental ruin but also have the potential to foster economic revitalization and social cohesion especially when cross-border cooperation and knowledge exchange become central to this endeavour.

However, it should not be lost that the complexity of sustainability challenges requires the pooling of resources, expertise, and experiences from across the globe. Therefore, international collaborations such as the Paris Agreement and the Convention on Biological Diversity which underscores the potential of collective action in addressing global environmental issues need to be emphasized. Further, for such collaborations to be effective, they must be underpinned by a commitment to equity and justice, ensuring that all nations, particularly those most vulnerable to environmental and social upheavals, have a voice in shaping the global sustainability agenda. This can be done by ensuring practical policies and regulations are put in place to safeguard and expedite the implementation, as well as force accountability (United Nations, 2022b).

A unified approach to tackling the intertwined challenges of environmental degradation, social injustice, and economic inequality is imperative. This approach calls for a reimagining of global governance structures, economic systems, and social norms to foster a world where sustainability is embedded in every facet of human endeavour (Allam, Bibri, Chabaud et al., 2022; Allam, Chabaud, et al., 2023). It necessitates a departure from siloed thinking and fragmented initiatives, advocating for an integrated strategy that leverages the strengths of ecological economics, social ecology, and environmental justice.

References

Agyeman, J., & Evans, B. (2004). 'Just Sustainability': The Emerging Discourse of Environmental Justice in Britain? *The Geographical Journal, 170*(2), 155–164.

Allam, M. Z. (2019). *Urban Resilience and Economic Equity in an Era of Global Climate Crisis.* University of Sydney,

Allam, Z. (2019). Enhancing Renewable Energy Adoption in Megacities through Energy Diversification, Land Fragmentation and Fiscal Mechanisms. *Sustainable Cities Society*, 101841.

Allam, Z. (2020a). The First 50 Days of COVID-19: A Detailed Chronological Timeline and Extensive Review of Literature Documenting the Pandemic. In *Surveying the COVID-19 Pandemic its Implications* (p. 1).

Allam, Z. (2020b). The Second 50 Days: A Detailed Chronological Timeline and Extensive Review of Literature Documenting the COVID-19 Pandemic From Day 50 to Day 100. In *Surveying the COVID-19 Pandemic its Implications* (p. 9).

Allam, Z. (2020c). *Surveying the Covid-19 Pandemic and Its Implications: Urban Health, Data Technology and Political Economy.* Elsevier.

Allam, Z. (2020d). The Third 50 Days: A Detailed Chronological Timeline and Extensive Review of Literature Documenting the COVID-19 Pandemic From Day 100 to Day 150. In *Surveying the Covid-19 Pandemic its Implications* (p. 41).

Allam, Z. (2020e). Vital COVID-19 Economic Stimulus Packages Pose a Challenge for Long-Term Environmental Sustainability. In *Surveying the Covid-19 Pandemic its Implications* (p. 97).

Allam, Z., Bibri, S. E., Chabaud, D., & Moreno, C. (2022). The '15-Minute City' Concept Can Shape a Net-Zero Urban Future. *Humanities and Social Sciences Communications, 9*(1), 126. https://doi.org/10.1057/s41599-022-01145-0

Allam, Z., Bibri, S. E., Jones, D. S., Chabaud, D., & Moreno, C. J. S. (2022b). Unpacking the '15-Minute City' via 6G, IoT, and Digital Twins: Towards a New Narrative for Increasing Urban Efficiency, Resilience, and Sustainability. *Sensors, 22*(4), 1369.

Allam, Z., Chabaud, D., Gall, C., Pratlong, F., & Moreno, C. (2023). On Proximity-Based Dimensions and Urban Planning: Historical Precepts to the 15-Minute City. In Z. Allam, D. Chabaud, C. Gall, F. Pratlong, & C. Moreno (Eds.), *Resilient and Sustainable Cities* (pp. 107–119). Elsevier.

Allam, Z., Jones, D., & Thondoo, M. (2020a). Decarbonization and Urban Sustainability. In *Cities and Climate Change* (pp. 33–54): Palgrave Macmillan.

Allam, Z., Jones, D., & Thondoo, M. (2020b). Economically Incentivizing Urban Sustainability and Resilience. In *Cities and Climate Change* (pp. 83–106). Palgrave Macmillan.

Allam, Z., & Jones, D. J. E. (2019). Climate Change and Economic Resilience through Urban and Cultural Heritage: The Case of Emerging Small Island Developing States Economies. *Economies, 7*(2), 62.

Allam, Z., Moreno, C., Chabaud, D., & Pratlong, F. (2022). Proximity-Based Planning and the "15-Minute City": A Sustainable Model for the City of the Future. In *The Palgrave Handbook of Global Sustainability*. Springer.

Allam, Z., Sharifi, A., Giurco, D., & Sharpe, S. A. (2022). Green New Deals Could Be the Answer to COP26's Deep Decarbonisation Needs. *Sustainable Horizons, 1*, 100006.

Allam, Z., Cheshmehzangi, A., & Jones, D. S. (2023). *Climate and Social Justice: The Political Economy of Urban Resilience and Mercantilism*. Springer.

Barry, J., & Quilley, S. (2009). The Transition to Sustainability: Transition Towns and Sustainable Communities. *Advances in Ecopolitics, 4*, 1–28. https://doi.org/10.1108/S2041-806X(2009)0000004004

Bartelmus, P. (1996). 6 Environmental Accounting: A Framework for Assessment and Policy Integration. In *Macroeconomics and the Environment*. International Monetary Fund.

Bookchin, M. (1982). *The Ecology of Freedom: The Emergence and Dissolution of Hierarchy*. Cheshire Books.

Bookchin, M. (2007). *Social Ecology and Communalism*. AK Press.

Boulton, C. A., Lenton, T. M., & Boers, N. (2022). Pronounced Loss of Amazon Rainforest Resilience Since the Early 2000s. *Nature Climate Change, 12*(3), 271–278. https://doi.org/10.1038/s41558-022-01287-8

Boyle, A. D., Leggat, G., Morikawa, L., Pappas, Y., & Stephens, J. C. (2021). Green New Deal Proposals: Comparing Emerging Transformational Climate

Policies at Multiple Scales. *Energy Research & Social Science, 81*, 102259. https://doi.org/10.1016/j.erss.2021.102259

Brown, T. C., Bergstrom, J. C., & Loomis, J. B. (2007). Defining, Valuing, and Providing Ecosystem Goods and Services. *Natural Resources Journal, 47*(2), 329–376.

Bullard, R. D. (1990). *Dumping in Dixie: Race, Class, and Environmental Quality*. Westview Press.

Calhoun, C., & Fong, B. Y. (2022). Introduction. In C. Calhoun & B. Y. Fong (Eds.), *The Green New Deal and the Future of Work* (pp. 1–20). Columbia University Press.

Cheshmehzangi, A. (2020). The Analysis of Global Warming Patterns from 1970s to 2010s. *Atmospheric and Climate Sciences, 10*(3). https://doi.org/10.4236/acs.2020.103022

Cheshmehzangi, A., & Dawodu, A. (2019). *Sustainable Urban Development in the Age of Climate Change—People: The Cure or Curse*. Palgrave Macmillan.

Cheshmehzangi, A., You, N., Siri, J., & Birch, E. (2024). *Harnessing Urban Innovation to Unlock the Sustainable Development Goals*. Springer.

Chowdhury, S. (2021, February 8). South Korea's Green New Dealinthe Year of Transition. https://www.undp.org/blogs/south-koreas-green-new-deal-year-transition

Cole, L. W., & Foster, S. R. (2001). *From the Ground Up: Environmental Racism and the Rise of the Environmental Justice Movement*. NYU Press.

Congress. (2021, April 19). H.R.2664—117th Congress (2021–2022): Green New Deal for Public Housing Act. https://www.congress.gov/bill/117th-congress/house-bill/2664/text

Dhunny, A., Allam, Z., Lobine, D., & Lollchund, M. J. E. (2019). Sustainable Renewable Energy Planning and Wind Farming Optimization from a Biodiversity Perspective. *Energy, 185*, 1282–1297.

Dolšak, N., & Prakash, A. (2018). The Politics of Climate Change Adaptation. *Annual Review of Environment and Resources, 43*(1), 317–341. https://doi.org/10.1146/annurev-environ-102017-025739

EPA. (1998, February 16). Summary of Executive Order 12898—Federal Actions to Address Environmental Justice in Minority Populations and Low-Income Populations. *Laws & Regulations*. https://www.epa.gov/laws-regulations/summary-executive-order-12898-federal-actions-address-environmental-justice#:~:text=Clinton%20in%201994.,environmental%20protection%20for%20all%20communities

European Commission. (2021). INTERREG: Cross-border, Transitional, and Interregional Cooperation in Europe. https://ec.europa.eu/regional_policy/en/policy/cooperation/european-territorial/interreg/

European Council. (2021). Fit for 55. https://www.consilium.europa.eu/en/policies/green-deal/fit-for-55-the-eu-plan-for-a-green-transition/#:~:text=Fit%20for%2055%20refers%20to,line%20with%20the%202030%20goal

Fischer, C., & Jacobsen, G. D. (2021). The Green New Deal and the Future of Carbon Pricing. *Journal of Policy Analysis and Management, 40*(3), 988–995. https://doi.org/10.1002/pam.22313

Gómez, R., Aguirre, J., Oliveros, L., Paladines, R., Ortiz, N., Encalada, D., & Armenteras, D. (2023). A Participatory Approach to Economic Valuation of Ecosystem Services in Andean Amazonia: Three Country Case Studies for Policy Planning. *Sustainability, 15*(6), 4788.

Government of Germany. (2020). Overveiw of Germany Resource Efficiency Programme (ProgRess). https://www.bmuv.de/en/topics/water-management/overview-resource-efficiency/overview-of-german-resource-efficiency-programme-progress

Hoehn, D., Laso, J., Margallo, M., Ruiz-Salmón, I., Amo-Setién, F. J., Abajas-Bustillo, R., Sarabia, C., Quiñones, A., Vázquez-Rowe, I., Bala, A., & Aldaco, R. (2021). Introducing a Degrowth Approach to the Circular Economy Policies of Food Production, and Food Loss and Waste Management: Towards a Circular Bioeconomy. *Sustainability, 13*(6), 3379.

Hummel, D., Jahn, T., Keil, F., Liehr, S., & Stieß, I. (2017). Social Ecology as Critical, Transdisciplinary Science—Conceptualizing, Analyzing and Shaping Societal Relations to Nature. *Sustainability, 9*(7), 1050.

Kish, K., & Farley, J. (2021). A Research Agenda for the Future of Ecological Economics by Emerging Scholars. *Sustainability, 13*(3), 1557.

Le, T.-A.T., Vodden, K., Wu, J., Bullock, R., & Sabau, G. (2024). Payments for Ecosystem Services Programs: A Global Review of Contributions Towards Sustainability. *Heliyon, 10*(1), e22361. https://doi.org/10.1016/j.heliyon.2023.e22361

Mahat, T. J., Bláha, L., Uprety, B., & Bittner, M. (2019). Climate Finance and Green Growth: Reconsidering Climate-Related Institutions, Investments, and Priorities in Nepal. *Environmental Sciences Europe, 31*(1), 46. https://doi.org/10.1186/s12302-019-0222-0

Midttun, A., Witoszek, N., Joly, C., Karlsson-Vinkhuyzen, S., Olsen, P., Olsson, L., Sigurjonssón, Þ.O., Toje, A., Trägardh, L., & Ostergard, U. (2011). The Nordic Model: Is It Sustainable and Exportable? *International Journal of Foresight and Innovation Policy.*

Natter, A. (2019, February 25). Alexandria Ocasio-Cortez's Green New Deal Could Cost $93 Trillion, Group Says. https://www.bloomberg.com/news/articles/2019-02-25/group-sees-ocasio-cortez-s-green-new-deal-costing-93-trillion

Neal, L. (2021). *Ecological Contradictions of Labour's Green New Deal*. https://www.ipe-berlin.org/fileadmin/institut-ipe/Dokumente/Working_Papers/ipe_working_paper_152.pdf

Obi, C. (2010). Environmental Rights, the Oil Industry, and the Niger Delta People's Struggle for Social Justice. In I. U. Asobele & I. A. Agbu (Eds.), *Law, State and Society in Modern Nigeria: Essays in Honor of the Late Chief Bola Ige, SAN, the Cicero of Esa Oke* (pp. 1–25). Nigeria University Press.

Pauli, B. J. (2020). The Flint Water Crisis. *Wiley Interdisciplinary Reviews: Water, 7*(3), e1420. https://doi.org/10.1002/wat2.1420

Portelli, M.-S., Conrad, E., & Galdies, C. (2020). Developing. *Sustainability, 12*(22), 9519.

Puaschunder, J. (2021). Equitable Green New Deal. *SSRN Electronic Journal*. https://doi.org/10.2139/ssrn.3926705

Pulido, L. (2000). Rethinking Environmental Racism: White Privilege and Urban Development in Southern California. *Annals of the Association of American Geographers, 90*(1), 12–40. https://doi.org/10.1111/0004-5608.00182

Rawtani, D., Gupta, G., Khatri, N., Rao, P. K., & Hussain, C. M. (2022). Environmental Damages Due to War in Ukraine: A Perspective. *Science of the Total Environment, 850*, 157932. https://doi.org/10.1016/j.scitotenv.2022.157932

Rechsteiner, R. (2021). German Energy Transition (Energiewende) and What Politicians Can Learn for Environmental And Climate Policy. *Clean Technologies and Environmental Policy, 23*(2), 305–342. https://doi.org/10.1007/s10098-020-01939-3

Rifkin, J. (2019). *The Green New Deal: Why the Fossil Fuel Civilization Will Collpase by 2028, and the Bold Economic Plan to Save Life on Earth*. St. Martin's Press.

Saikia, T., Hazarika, J., Kurmi, K. K., & Gogoi, D. (2023). Seed Bomb: A New Approaches of Afforestation—An Overview. *The Pharma Innovation Journal, 12*(10), 2114–2118. https://doi.org/10.22271/tpi

Sidiropoulos, E. (2022). Africa; Aspiring to Greater Global Agency. https://carnegieeurope.eu/2022/02/17/africa-aspiring-to-greater-global-agency-pub-86371

Spash, C. L. (2008). How Much Is that Ecosystem in the Window? The One with the Bio-diverse Trail. *Environmental Values, 17*(2), 259–284.

Strielkowski, W., Civín, L., Tarkhanova, E., Tvaronavičienė, M., & Petrenko, Y. (2021). Renewable Energy in the Sustainable Development of Electrical Power Sector: A Review. *Energies, 14*(24), 8240.

Tan, Z., Wu, Y., Gu, Y., Liu, T., Wang, W., & Liu, X. (2022). An Overview on Implementation of Environmental Tax and Related Economic Instruments in Typical Countries. *Journal of Cleaner Production, 330*, 129688. https://doi.org/10.1016/j.jclepro.2021.129688

The World Bank. (2023). *From Evolution to Revolution: 10 Years of Green Bond Event*. https://treasury.worldbank.org/en/about/unit/treasury/ibrd/ibrd-green-bonds

UNFCCC. (2020, November 4). Joint Statement on the US Withdrawal From the Paris Agreement. https://unfccc.int/news/joint-statement-on-the-us-withdrawal-from-the-paris-agreement

United Nations. (2017). *New Urban Agenda*. https://uploads.habitat3.org/hb3/NUA-English.pdf

United Nations. (2020). United Nations Sustainable Development Goals. https://sdgs.un.org/goals

United Nations. (2022a). *A Just Green Transition: Concepts and Practice So Far*. https://www.un.org/development/desa/dpad/wp-content/uploads/sites/45/publication/PB_141.pdf

United Nations. (2022b, June 2). Security Council Must Strengthen, Uphold Responsibility to Ensure Accountability When International Law Is Violated, Speakers Stress in All-Day Debate. *9052nd Meeting*. https://press.un.org/en/2022/sc14919.doc.htm

United Nations. (2023a, October 16). Geopolitical Conflicts, Economic Uncertainty Darkening Sustainable Development Prospects, Officials Warn, as Second Committee Takes Up Regional Commissions. *Seventy-Eight Session*. https://press.un.org/en/2023/gaef3592.doc.htm

United Nations. (2023b). Payments for Environmental Services Program—Costa Rica. https://unfccc.int/climate-action/momentum-for-change/financing-for-climate-friendly-investment/payments-for-environmental-services-program#:~:text=Costa%20Rica's%20Payments%20for%20Environmental,the%20country%20and%20the%20region

United Nations Development Programme (UNDP). (2015). *Sustainable Development Goals*. UNDP.

United Nations Environment Programme. (2023, February 22). A Green Wall to Promote Peace and Restore Nature in Africa's Sahel Region. https://www.unep.org/news-and-stories/story/green-wall-promote-peace-and-restore-nature-africas-sahel-region

United Nations Framework Convention on Climate Change. (2015). Paris Agreement. https://unfccc.int/sites/default/files/english_paris_agreement.pdf

Ura, K., Alkire, S., Zangmo, T., & Wangdi, K. (2012). *An Extensive Analysis of GNH Index*. https://ophi.org.uk/wp-content/uploads/Ura_et_al_Extensive_analysis_of_GNH_index_2012.pdf

Van Horne, Y. O., Alcala, C. S., Peltier, R. E., Quintana, P. J. E., Seto, E., Gonzales, M., & Beamer, P. I. (2023a). An Applied Environmental Justice

Framework for Exposure Science. *Journal of Exposure Science & Environmental Epidemiology, 33*(1), 1–11. https://doi.org/10.1038/s41370-022-00422-z

Venturini, F., Değirmenci, E., & Morales Bernardos, I. (2019). *Social Ecology and the Right to the City : Towards Ecological and Democratic Cities*.

Walker, R. A. (2022). From the New Deal to the Green New Deal. In C. Calhoun & B. Y. Fong (Eds.), *The Green New Deal and the Future of Work* (pp. 23–52). Columbia University Press.

Wilson, M. (2019). The Green Economy the Dangerous Path of Nature Commoditization. *Consilience, 21*, 86–99.

Wironen, M. B., & Erickson, J. D. (2020). A Critically Modern Ecological Economics for the Anthropocene. *The Anthropocene Review, 7*(1), 62–76. https://doi.org/10.1177/2053019619884485

World Economic Forum. (2023, March 17). Why Industry Collaboration Is Necessary for a Sustainable Future. *Climate and Nature*. https://www.weforum.org/agenda/2023/03/radical-collaboration-for-a-sustainable-future-the-case-for-sustainability-collaboration/

Yu, V. P. (2023, December 13). Addressing the Climate Technology Gap in Developing Countries Through Effective Technology Transfer. https://tesforum.org/latest/addressing-the-climate-technology-gap-in-developing-countries-through-effective-technology-transfer

Zaheer, A., & Jones, D. S. (2018). *Promoting Resilience, Liveability and Sustainability Through Landscape Architectural Design: A Conceptual Framework for Port Louis, Mauritius; a Small Island Developing State*. Paper presented at the IFLA 2018: Biophilic City, Smart Nation, and Future Resilience: Proceedings of the 55th International Federation of Landscape Architects World Congress 2018.

Zhou, S., & Peng, F. (2023). The Impact of Technology Transfer on the Green Innovation Efficiency of Chinese High-Tech Industry. *Frontiers in Sociology, 8*. https://doi.org/10.3389/fsoc.2023.1141616

Zolyomi, A., Franklin, A., Smith, B., & Soliev, I. (2023). Ecosystem Services As the Silver Bullet? A Systematic Review of How Ecosystem Services Assessments Impact Biodiversity Prioritisation in Policy. *Earth System Governance, 16*, 100178. https://doi.org/10.1016/j.esg.2023.100178

CHAPTER 3

Technological Innovation and Sustainable Transitions

Abstract Technological innovations play a critical role in facilitating sustainable transitions in the face of contemporary climate challenges. Key advancements in renewable energy, smart grids, and circular economy practices are essential for sustainable urban development. Since the Brundtland Commission's seminal report, the sustainability discourse has expanded to encompass economic viability and social equity alongside environmental concerns. Renewable energy technologies, including solar, wind, and bioenergy, present viable alternatives to fossil fuels, significantly reducing carbon emissions and enhancing energy security. Smart grids, through the dynamic management of energy supply and demand, leverage real-time data to optimize energy efficiency and reliability. The circular economy reimagines growth and consumption patterns, emphasizing the reuse, refurbishment, and recycling of resources to minimize waste. Despite the promising potential of these technologies, their integration faces significant challenges, such as economic constraints, technological limitations, and regulatory hurdles. Addressing these obstacles requires a multifaceted and collaborative approach, involving governments, businesses, civil society, and the academic community. This chapter delves into the intricacies of these technological innovations, exploring their roles as pillars of sustainable transitions and offering insights into overcoming the barriers to their adoption. The collective effort to integrate these technologies is crucial for advancing towards a more sustainable, resilient, and equitable urban future.

Z. Allam and A. Cheshmehzangi, *Sustainable Futures and Green New Deals*, https://doi.org/10.1007/978-3-031-63642-4_3

Keywords Technological Innovation · Sustainable Transitions · Renewable Energy · Smart Grids · Circular Economy · Urban Development · Energy Efficiency · Climate Change

1 Introduction

The dawn of the Green New Deal (GND) and the continuing discourse around sustainability more so on urban transitions underscore the critical juncture at which modern society finds itself amid numerous climate change-related challenges (Allam et al., 2022b). Among the strategies being pursued is the adoption of technological innovation to serve as a cornerstone of the desired transition (Cheshmehzangi, 2020; Cheshmehzangi et al., 2021), particularly in renewable energy generation (Barragán-Escandón et al., 2017; Allam et al., 2021) and sustainable transportation. It has widely been acknowledged that technological advancements serve not merely as facilitators but as catalysts for sustainable development, intertwining with economic, environmental, and social fibres to weave a tapestry of societal advancement. This chapter will delve into the intricacies of renewable energy, smart grids, and circular economy practices, and elucidate their roles as pillars of sustainable transitions, especially in this era when there is increased focus on GND.

Since the Brundtland Commission's 'Our Common Future' report in 1987 introduced the concept of sustainable development (Burundtland, 1987), the dialogue on sustainability has evolved from a mere environmental concern to a comprehensive framework incorporating economic viability and social equity. Technological innovations in renewable energy (International Renewable Energy Agency, 2020; Allam, 2019), smart grid technologies (Kabeyi & Olanrewaju, 2023), and circular economy (Awan & Sroufe, 2022) practices epitomize this shift, offering a trinity of solutions to the multifaceted challenges of sustainable urban development. These innovations herald a new era of energy production, consumption, and waste management, promising significant strides towards reducing carbon emissions, enhancing energy efficiency, and fostering a regenerative economic model that decouples growth from resource consumption (UNCTAD, 2023). Solar, wind (Dhunny et al., 2019), hydroelectric, and other renewable sources such as bioenergy offer promising alternatives to the perennial challenges of fossil fuel dependency (Strielkowski et al., 2021). On the other hand, smart grids are expected to enable dynamic management of energy supply and demand, especially by relying on real-time data analytics and intelligent control systems. This way, the energy sector would become more efficient, reliable, and sustainable, and in the long run, would benefit everyone in the society despite their social status or economic standing (Mostafa et al., 2022). The circular economy on its part is expected to transform the urban development agendas in a sustainable way; hence, redefining concepts of growth, productivity, and consumption (Awan & Sroufe, 2022).

The efforts to integrate these technologies into the fabric of urban life have always encountered unprecedented hurdles and challenges. Such include economic constraints, technological limitations, and regulatory hurdles that often derail the full adoption of sustainable technologies (Allam and Dhunny, 2019; Allam, 2021). Moreover, the transition towards sustainability has and continues to necessitate a paradigm shift in societal values, consumer behaviours, and business models for it to succeed. This then calls for a collaborative approach across various sectors of society, governments, businesses, civil society organizations, and the academic community. As expressed by Martínez-Peláez et al. (2023), these stakeholders need to converge in a concerted effort to drive the adoption of sustainable technologies, crafting policies, and fostering an ecosystem that nurtures innovation and accelerates the shift towards sustainable urban futures.

The emergence and acceptability of renewable energy technologies marks a pivotal shift in the quest for a sustainable energy future. Bioenergy, solar, hydroelectric, and wind, sources serve as welcome relief to the dominance of fossil fuel, thus, highlighting the potential for decarbonized energy systems. The integration of renewable energy into urban energy networks not only helps reduce greenhouse gas emissions but also helps in diversifying the energy mix as well as enhancing energy security. These alternative sources are becoming reliable in contributing to economic resilience and environmental sustainability, as well as bridging the societal energy distribution, particularly by fostering the last-mile connectivity initiatives in different parts of the world.

Despite the numerous benefits, renewable energy technologies deployment across different parts of the world are encumbered by political, economic, and infrastructural challenges, particularly in the global south and less developed countries. The initial cost of renewable energy systems, coupled with the need for substantial investments in grid infrastructure to accommodate variable energy sources, poses significant hurdles, thus making fossil fuel power energy to continue fashionable (Nadeem et al., 2023). However, it should be noted that this was not the case pre-COVID-19, as initial costs had started to drop significantly, especially for solar power. After the pandemic, factors such as supply chain challenges prompted the cost of basic materials such as steel, silicon, and others to rise, leading to a rise in components such as solar panels and storage batteries (Catalbas, 2022). For instance, in 2024, the IEA report that the cost of steel basic for utility-scale PV and onshore wind plants rose by 75% in China, 160% in the United States, and approximately 270% in the Euro region (IEA, 2023b). Despite these challenges, the long-term economic benefits of renewables, including job creation, energy cost savings, and the stimulation of green industries, underscore the imperative for strategic investments in renewable energy infrastructure. For instance, it is reported that there are over 35 million jobs created in the renewable energy sector, with 4.7 million of those having been created between 2019 and 2023. Such were due to the negative impacts of COVID-19 on the fossil fuel jobs, where to date, the sector has not yet recovered approximately 1.3 million jobs (IEA, 2023c).

Smart grids on their part represent the link between information technology and energy infrastructure, enabling active management of energy supply and demand. Through real-time data analytics and intelligent control systems, smart grids enhance and allow for a seamless flow of both electricity and data between consumers and their suppliers; thus, facilitating efficiency, reliability, security, and sustainability of urban energy systems (Gupta & Chaturvedi, 2023). This potential is further amplified when renewable energy sources are integrated with smart grids. This then facilitates a seamless transition to a decentralized energy paradigm.

However, just like the renewable energy generation, the economic and environmental dividends of smart grids are hampered by numerous challenges, particularly those associated with their technological and regulatory demands (Kabeyi & Olanrewaju, 2023). Among the most popular challenges include cybersecurity concerns, the complexity of grid modernization, and the need for regulatory frameworks that incentivize innovation. These challenges are experienced even in developed countries, where one would imagine a smooth implementation. For instance, in Germany, challenges of data privacy regulations and lack of standardizations derail adoption, while in the United States, there are no consistent regulations or policies to guide successful implementation (Moreno Escobar et al., 2021). For developing economies, cases such as insufficient funding, low technical expertise, and a low number of skilled individuals to undertake jobs remain a key challenge. The role of smart grids in advancing sustainable urban transitions is incontrovertible, necessitating a collaborative approach to surmount the barriers to their adoption.

In the recent past, the concept of circular economy has emerged as among the transformative models for sustainable urban development that could help countries navigate the environmental sustainability, social justice, and economic development terrain. This concept is redefining concepts of growth, productivity, and consumption through its emphasis on the reuse, refurbishment, and recycling of materials. This way, it has brought about practical ways of minimizing waste and optimizing resource efficiency; hence, fostering a regenerative economic system (Hoehn et al., 2021). For the circularity aspect to be achieved, technological innovation is necessary, from advanced recycling technologies to platforms for sharing and reusing resources. There is also a need to address perennial negativities such as cultural and corporate inertia, regulatory gaps, and market dynamics that favour linear consumption models at the expense of circularity (Yamoah et al., 2022). Overcoming these barriers and guaranteeing the continuity of this concept requires systemic change, underscored by policy innovation, business model transformation, and shifts in consumer behaviour.

Overall, the transition to sustainable urban futures can be argued to be laden with myriad challenges, from economic and technological barriers to regulatory and societal hurdles. These calls for a multifaceted approach to tackle, including a blending of policy interventions, financial incentives, technological innovation, and extension of support to economies grappling with financial challenges.

The collective endeavour to foster sustainable urban transitions hinges on the active engagement of all societal sectors. Businesses must embrace sustainability as a core strategic imperative, driving innovation and adopting sustainable practices. Governments play a crucial role in shaping policies and providing the regulatory and financial frameworks that support sustainable development. Civil society organizations and the academic community are pivotal in advocating for change, raising public awareness, and contributing to the knowledge base on sustainable technologies. Through collaborative efforts, the multifarious challenges of sustainable transitions can be surmounted, paving the way for a sustainable, resilient, and equitable urban future.

2 Renewable Energy Technologies

The landscape of renewable energy has witnessed significant evolution over the past decades, particularly after the break of the fourth industrial revolution. As a result, the sector is no longer deemed transitional peripheral alternatives but is now a central pillar of the global energy matrix. Substantial advancements have been achieved in solar, wind, hydro, and bioenergy technologies which are now major drivers in the energy sector's transformative journey, each contributing uniquely to the renewable energy portfolio. The growth of these technologies is not merely reflective of technological advancements but also of a collective global commitment to sustainable energy futures, brought about by fate (climate change), and the increasing internal demand to provide for the increasing local citizenry. The International Renewable Energy Agency (IRENA) reports a consistent increase in renewable energy capacity, underscoring the pivotal role of these technologies in achieving energy transition goals (International Renewable Energy Agency, 2020).

In the discourse about climate action, renewable energy technologies have become central due to their ability and potential to offer viable means to reduce global carbon footprints and diminish dependency on fossil fuels. They also offer cheaper alternatives to increasing global fuel prices prompted by issues like geopolitical conflicts, supply chain problems, and business-oriented approaches by key fossil fuel-producing countries.

The world is endowed with diverse natural resources that could be maximized to make these renewable technologies produce sufficient energy with minimal to zero emissions. According to IRENA, available resources have the potential to produce more than 90% of the world's electricity demand, and the projection is that by 2050, this needs to be realized. The potential of global natural resources in this endeavour has been rightly acknowledged in the Paris Agreement. With the global temperatures increasing significantly, almost by 2°C as highlighted in a report by IPCC (IPCC, 2023), the shift from fossil fuels to renewable energy source is thus not only a strategic manoeuvre to enhance energy security and reduce the volatility associated with fossil fuel markets but also an environmental imperative to combat the climate change challenges. The economic implications of renewable energy technologies extend far beyond their environmental benefits, catalysing job creation (International Renewable Energy Agency, 2020), reducing energy costs, and opening new avenues for investment. The renewable energy sector has proven to be more labour-intensive than the fossil fuel industry (employing over 35 million jobs compared to 32 million by fossil fuel industry as of 2023) (IEA, 2023c), implying that the transition towards renewables is not only environmentally prudent but also economically beneficial. Furthermore, the decreasing cost of renewable energy technologies, particularly solar and wind, has rendered them increasingly competitive against traditional energy sources, driving down energy costs for consumers and businesses alike (Yamoah et al., 2022).

Despite the promising outlook, the adoption of renewable energy technologies faces several challenges. Infrastructure inadequacies particularly in developing countries, the need for substantial initial investments (which increased substantially due to the impacts of the COVID-19 pandemic as discussed above), and the complexities of energy storage represent significant hurdles. The intermittent nature of solar and wind energy, for instance, necessitates sophisticated storage solutions to ensure a reliable energy supply, presenting technical and economic challenges that must be addressed to fully leverage the potential of renewable energy technologies (Pollitt, 2012). In particular, the costs of advanced technologies need to be deployed in the manufacturing of batteries, more so with an aim to lower the costs which are substantially higher (Catalbas, 2022). Countries like Denmark and Germany have demonstrated leadership in integrating renewable energy into their national grids, providing lessons on policy frameworks, technological innovation, and public–private partnerships that can facilitate the transition to renewable energy. These examples underscore the importance of cohesive strategies and adaptive measures in overcoming the complexities of renewable energy adoption (Lipp, 2007).

The future of renewable energy technologies is marked by both challenges and opportunities, with emerging innovations poised to further revolutionize the energy landscape. It should be acknowledged that despite the challenges, substantial advancements have been made, especially in making renewables, almost as competitive as fossil fuels, and were it not for the impacts of the COVID-19 pandemic, a different report would be available. Advances in energy storage, such as battery technologies and hydrogen fuel, alongside the integration of artificial intelligence and blockchain in energy systems, promise to enhance the efficiency and reliability of renewable energy. As the world continues to grapple with the exigencies of climate change and energy security, fossil fuel shortage in some countries among other issues, the trajectory of renewable energy technologies remains the most potent new normal, signalling a sustainable and resilient energy future.

3 Smart Grids and Energy Efficiency

Smart grids represent a transformative evolution in the domain of energy management, signifying a shift from traditional, unidirectional energy distribution systems to dynamic, interactive, interconnected networks capable of real-time monitoring, analysis, and control. They represent advanced, modern electricity networks that utilizes digital technologies, software, and sensors to manage the balance between supply and demand in real time (IEA, 2023a). They have become very practical in addressing the complexities of modern energy demands and in facilitating the integration of renewable energy sources, thus enhancing the reliability and efficiency of electricity distribution. The responsive nature of smart grids aligns seamlessly with global initiatives such as the Sustainable Development Goals (e.g. SDG 7 which emphasizes clean and affordable energy [United Nations, 2020]), aimed at mitigating climate change by reducing greenhouse gas emissions (Cheshmehzangi & Dawodu, 2019; Cheshmehzangi et al., 2024). Additionally, they help empower nations towards energy independence, breaking free from the constraints of conventional fuel sources (Kabeyi & Olanrewaju, 2023).

The contribution of smart grids to energy efficiency, reliability, and sustainability is multifaceted. For instance, regarding energy efficiency, they support demand response programs (e.g. the use of smart thermostats), where consumers are able to adjust their energy use based on real-time data such as pricing signals, and peak demand; hence, contributing to energy saving. The use of technologies such as sensors and machine learning algorithms helps in real-time data acquisition and analysis; thus allowing for automated actions such as adjustments to voltage power flow and efficient equipment usage. The resulting outcomes include optimized systems, reduced energy wastage, and enhanced system efficiency. Their ability to anticipate and respond to issues in the grid in real-time enhances reliability (e.g. self-healing grids and predictive maintenance), minimizing downtime and the potential for power outages (Mahmoud et al., 2021). Furthermore, by facilitating the integration of variable renewable energy sources, smart grids are pivotal in the transition towards more sustainable energy systems, ensuring that the maximum possible amount of renewable energy can be utilized to meet demand.

In addition, the integration of the Internet of Things (IoT) and Artificial Intelligence (AI) technologies within smart grids plays a significant role in modern energy management endeavours. IoT devices provide a network of connected sensors and metres that gather vast amounts of data on energy usage, grid performance, and environmental conditions (Kirmani et al., 2023; Renugadevi et al., 2023). AI algorithms process this data, enabling predictive maintenance, demand forecasting, and anomaly detection, thereby enhancing grid performance and facilitating the seamless integration of renewable energy sources (Elabbassi et al., 2024). This synergy between IoT and AI becomes handy also in accelerating the adoption of renewable energy by optimizing its integration and distribution.

Economically, smart grids can lead to significant cost savings for utilities and consumers through improved energy efficiency and reduced need for peak power generation capacity. A real-life example to affirm this is Texas' Smart Grid Initiative which began in 2009 aiming at modernizing the electricity grid and optimize its performance in the state of Texas (Austin Energy, 2019). Environmentally, smart grids enhance the integration of renewable energy sources and the optimization of energy distribution leading to lower greenhouse gas emissions and a reduced carbon footprint. The transition to smart grids is thus not only a strategic investment in future energy security but also a commitment to environmental stewardship and sustainable development.

Like any other modern technology, the implementation of smart grids has and continues to face diverse challenges and barriers. However, cybersecurity emerges as a critical concern, given the reliance of smart grids on digital communication networks, making them vulnerable to hacking and cyber-attacks. A case in point is the cyberattack on the Ukraine power system in 2015 and 2016 that left over 1 million people without power for more than 24 hours, with disruption on businesses, hospitals, and essential services causing immense economic losses (Whitehead et al., 2017).

Besides cybercrime, privacy issues are also live particularly arising from the detailed consumption data collected, necessitating robust data protection measures. This could be well explained by reviewing the German court ruling on smart metre data in 2020, arguing that the collection and storage of energy consumption data violated individual privacy rights. This was necessitated by a lack of policy to safeguard to privacy of consumers and a lack of transparency and control (Sylla & Springmann, 2023). Additionally, the technological complexity and the need for substantial infrastructure investment pose significant barriers to widespread adoption.

Overcoming these challenges requires a concerted effort from policymakers, industry stakeholders, and the public to foster a secure, privacy-respecting, and technologically advanced smart grid ecosystem (de Reuver et al., 2016). Around the world, numerous smart grid initiatives showcase the transformative potential of this technology. From Europe's ambitious Smart Grid Task Force to Singapore's Intelligent Energy System pilot, these projects demonstrate the global commitment to harnessing smart grid technology for sustainable energy management. The results, including reduced energy consumption, improved system reliability, and greater incorporation of renewable energy, highlight the critical role smart grids play in the global transition towards more sustainable and efficient energy systems (IEA, 2023a).

4 Circular Economy Practices

In the wake of initiatives such as the Green New Deal (GND), innovative economic approaches have emerged. Among these is the circular economy which presents a compelling solution to many of the sustainability challenges faced by contemporary society. It reimagines the economy as not just a closed loop, but a regenerative ecosystem, with metabolic processes that optimize resource use while also minimizing waste to guarantee the sustained health of the entire system (Fehrer et al., 2024). It signifies a departure from the entrenched linear economic model characterized by a take-make-dispose approach, advocating instead for a system where resources are utilized to their fullest potential and wastes are systematically designed out of the process (Ellen Macarthur Foundation, 2023). This model not only supports sustainable development but also promotes economic growth and innovation in a manner that is environmentally responsible and socially inclusive. Its promising benefits and environmental strengths have made it attractive in many parts of the world, particularly in developed economies that have been struggling to guarantee environmental sustainability and the creation of new jobs for their populace. Since the adoption of this concept, the European region for instance has managed to create approximately 4 million jobs, contributing over €126 billion to the EU economy as of 2019 report (Sabbati, 2021).

In the recent past, the integration of technology into circular economy practices has been of great significance, particularly in enabling timely and efficient tracking, recycling, and repurposing of resources. With technology, innovations such as advanced recycling processes break down materials that would have otherwise been discarded, transforming waste into valuable inputs for new products has become a mainstay. For instance, the deployment of IoT-enabled devices such as sensors has extended this capability by connecting products and materials in a network that enables the tracking of their usage and lifecycle, facilitating efficient resource management (Ding et al., 2023).

Another technology that has proven vital in the pursuit of circularity is Artificial Intelligence (AI) among other things further enhances circular economy processes by analysing data to optimize manufacturing and recycling operations and predicting when materials can be recovered and reused (Pathan et al., 2023; Sjödin et al., 2023). These technologies collectively ensure that products and materials are not only used more efficiently but also remain in the economic system for as long as possible, thereby reducing excessive extraction and consumption of ecological resources and minimizing waste.

At the economic development and growth frontier, the circular economy presents a myriad of benefits, chief among them being increased resource efficiency which directly translates to cost reductions. For instance, by adopting the circular economy model, Levi Strauss & Co. fashion company has managed to reduce its water consumption by approximately 96%, and this has translated to a reduction in the production costs of the company's products. The company has reported that over 80% of its product are now being made from 'water < less' technology (Levi Strauss & Co., 2019), and as such, they are able to optimize their profit margin, while remaining true to sustainability agendas.

By rethinking and redesigning products and processes to maximize the reuse of materials, businesses can significantly decrease their input costs. This model also spurs innovation and competitiveness by encouraging the development of new, sustainable methods of product design, production, and service delivery as showcased in the case of Levi Strauss & Co. Moreover, the transition towards a circular economy opens up new market opportunities and creates jobs, especially in the recycling, remanufacturing, and circular supply chain sectors. These economic incentives are powerful drivers for companies to adopt circular economy principles thus, reduce their affinity to the consumption of finite raw materials and reduce environmental externalities.

On the environmental front, the benefits of adopting a circular economy are profound and immeasurable. One key advantage that this model brings is prioritizing the reduction of waste at all stages of product manufacturing and encouraging the reuse and recycling of materials. As a result, significant amounts of greenhouse gas emissions are reduced, while at the same time, minimizing the extraction of raw materials, thus alleviating pressure on natural resources. A case in point to support this is the Ellen MacArthur Foundation's New Plastics Economy Initiative which was launched in 2016 with the aim of tackling the plastic pollution crisis. The target was to ensure that over 95% of plastic packaging materials that are lost to the economy after a short first-use cycle are maintained longer; hence, saving the global economy a supposedly loss of between $80-$120 billion each year (Ellen Macarthur Foundation, 2016). Furthermore, the circular economy contributes to pollution reduction and biodiversity conservation by ensuring that natural systems are not only protected but also restored and regenerated. This holistic approach to environmental stewardship is crucial for combating climate change and ensuring the long-term health and viability of our planet.

Despite the clear advantages, the shift towards a circular economy is impeded by several obstacles. One key impediment is cultural norms and consumer behaviour. From the literature, it has been established that cultural norms are deeply ingrained in the convenience and disposability of the linear economy hence, pose significant challenges in convincing people to embrace the circularity practices. For instance, with the dominance of 'fast fashion' and the culture of newness characterized by rapid trend cycles and low-cost clothing, consumers tend to constantly seek newness and have no room for concentrating or focusing on product longevity, repair or reuse (Andreadakis & Owusu-Wiredu, 2023). Such trends could be countered by adopting Patagonia's approach, where the commitment to recycling worn wear and investing in sustainable materials exemplifies how companies can lead the way in environmental stewardship while remaining profitable (Shourkaei et al., 2024).

Another example is the culture of associating new technologies with status; hence, a majority of devices such as electronics that could be maintained longer in the system are deemed obsolete only after a short period when a supposedly upgraded version is introduced (Oraee et al., 2024). The practical application of circular economy principles across various industries as has been witnessed in the recent past highlights the model's versatility and effectiveness, beyond the conventional norm and perspectives. For instance, as a counter to the planned obsolesce notion, it could be possible to adopt Fairphone's efforts in the electronics industry to produce easily repairable and upgradable smartphones (Fairphone, 2024). In the construction sector's exploration of recycled materials and modular construction techniques could significantly help in waste reduction and sustainability improvements in one of the most resource-intensive industries.

In most economies, some of the challenges can be attributed to the laxity in regulatory and policy frameworks which often lag behind the innovative practices of the circular economy. For instance, most have not incorporated or imagined the necessary incentives and support mechanisms to facilitate circular economy practices. As such, industry players are not motivated to move quickly and embrace the transition, as a majority would fancy the status quo. Among areas that governments and regulators can offer incentives and support is the adoption and acquisition of technological and relevant infrastructure. This could be through financial support to promote investments required to implement circular economy practices. Further, overcoming the above barriers necessitates a comprehensive strategy that includes policy innovation, cultural change, and the development of financial mechanisms that make circular economy practices both viable and attractive.

5 Overcoming Challenges and Barriers

The pursuit of sustainable urban transitions has been acknowledged to be riddled with complex challenges that impede the swift adoption and implementation of essential technological innovations. These challenges range from financial constraints, regulatory problems (stemming from governance (Allam, 2020b) and collaboration shortcomings), and technological issues. In order to succeed in urban transition, it is critical to conduct a thorough exploration of each of these obstacles and craft a viable solution.

In perspective, transitioning towards sustainable urban requires significant upfront investments, particularly if the prerequisite projects are to be built from scratch. Even where some form of infrastructure exists and requires to be upgraded or changed, huge financial resources will be required. For instance, it is estimated that for cities across the globe to transition to a more efficient transport system, a minimum incremental investment of $8.6 trillion will be required (Lazer et al., 2020). As such, responsible entities, organizations, or governments are required to mobilize the required financial support, but in most cases, accessing either public or private financial instruments and attracting potential investors is never easy for many cities. This is particularly true when those substantial initial investments are juxtaposed with the ambiguity surrounding financial returns. For instance, in relation to the example above, it is estimated that the initial return from the investment would come eight years later, with an annual return of about $320 billion. It is only by 2050 that the returns would exceed $1 trillion, hence, affirming that the break-even point would take a substantial number of years (Lazer et al., 2020). Despite the associated benefits, the time taken for return on investment (RoI) to be accrued may discourage some from investing. This situation is exacerbated by prevailing economic frameworks and market dynamics, which tend to favour traditional technologies over their sustainable counterparts, thereby relegating the latter to a less competitive standing. They also tend to favour developed economies, which might have some capacities or abilities to have RoI quicker than cities in less developed economies.

To navigate the financial hurdles, a multifaceted approach involving incentives, subsidies, and the introduction of innovative funding mechanisms is indispensable. Incentives and subsidies serve to lighten the financial burden associated with initial investments, rendering sustainable technologies more appealing to investors and consumers alike (Allam et al., 2020; Allam et al., 2023b). Furthermore, the advent of green bonds (The World Bank, 2023), sustainability-linked loans, and other novel financial instruments furnish the requisite capital for ambitious sustainable projects, thereby addressing the critical need for funding in the pursuit of sustainability.

It has been established that regulatory and policy frameworks in most countries frequently fail to keep pace with rapid technological advancements, engendering a discord that stifles the adoption of sustainable technologies (Allam et al., 2022a). That is, the rapid innovations in sectors like renewable energy, alternative mobility, and new sustainable building technologies are not backed by up-to-date policies or regulations (Allam et al., 2023c; Allam and Newman, 2023). In most cases, the policies are outdated, do not offer any incentives, do not guarantee privacy and security, or lack enforcement mechanisms. The obscure nature of regulatory landscapes, marked by complexity and fragmentation, compounds compliance difficulties, thereby stalling or discouraging innovation altogether (Salet & de Vries, 2019). In this context, the role of government is instrumental, not only in formulating new or updating existing policies but also in ensuring compliance and holding accountable stakeholders who fraught them. By fostering regulatory environments that incentivize investments in sustainable technologies and simplify their market introduction, governments can significantly mitigate these challenges. Measures such as the simplification of regulatory procedures, the provision of tax incentives for green investments, and the establishment of explicit sustainability benchmarks serve to guide industrial practices towards sustainability goals (Jordan & Lenschow, 2010).

The absence of, or obscure standardization and interoperability among sustainable technologies constitutes a notable impediment, particularly as major players tend to work in siloes to protect their property rights, or to yield maximum competitive edge. The lack of universally recognized standards complicates the integration of new technologies into pre-existing systems, leading to inflated costs and hampering their broad-based adoption (Choi et al., 2020). Addressing this barrier demands concerted efforts and bipartisan collaboration toward promoting technological standardization and interoperability. Standardization and certification bodies and industry consortia are pivotal and need to be supported, as they are tasked with formulating and enforcing standards that guarantee compatibility, thereby enabling the fluid integration of sustainable technologies across various sectors and platforms (Castka, 2020).

Cultural and social barriers, characterized by resistance to change and a lack of awareness regarding the capabilities of sustainable technologies, further impede progress (Allam et al., 2023a). Likewise, tackling these barriers demands an emphasis on approaches like supporting the education sector and promoting public awareness initiatives. Educational programs dedicated to sustainability and the significance of technological innovation play a vital role in cultivating a culture attuned to environmental stewardship and innovation (Pauw et al., 2015). Concurrently, public awareness campaigns that interpret the advantages of sustainable technologies will need to be effectively discussed to dispel misconceptions and alter perceptions, thereby fostering broader acceptance and uptake.

Further, in view of the inherent global nature of sustainability challenges, there is a need for international cooperation and partnerships. The solitary efforts of individual countries or entities are insufficient in addressing these pervasive challenges. International collaboration facilitates the exchange of knowledge, technology, and best practices, while partnerships spanning governments, the private sector, and civil society galvanize the resources and expertise required to confront sustainability challenges on a worldwide scale.

5.1 *The Role of Stakeholders in Accelerating Adoption*

The recent global challenges such as the COVID-19 pandemic compounded the need for unity and a non-partisan approach to tackling universal challenges (Allam, 2020a; Allam, 2020c). This unity cannot be overemphasized in the pursuit of transition to sustainable urban environments following the numerous challenges such as erratic weather conditions, droughts, health challenges, and many others. The participation of everyone is critical in fostering the adoption of sustainable technologies and practices. From businesses and industries to governments, civil society, the scientific community, international organizations, and public–private partnerships, the concerted actions of these actors are crucial in driving innovation, formulating supportive policies, raising public awareness, and facilitating knowledge exchange on a global scale.

Over the years, businesses and industries have proven to be indispensable regarding sustainable innovation. It is these entities that have been at the forefront of championing corporate sustainability strategies and committing sizeable investments to shape the development and deployment of green technologies. It is true that by integrating sustainability into their core business models, companies not only enhance their competitive profiles but also contribute to environmental conservation agendas and social well-being. Such strategies often include reducing energy consumption, minimizing waste, investing in renewable energy sources (Allam et al., 2023c; Chabaud et al., 2023), emphasizing corporate social responsibilities and fostering transparency and accountability. A practical case to elucidate this is *Unilever's Sustainable Living Plan* launched in 2010 which focus on improving health and well-being, halving the environmental impact of their products, and improving their value chain to better the livelihoods of global citizens. For instance, the company is determined to reduce its plastic packaging by adopting 100% recycled plastic by 2025 (Unilever, 2021).

The critical role that governments across the board play in creating the enabling environments necessary for sustainable transitions through the formulation of policies, regulations, and incentives cannot be overlooked. Over the years, governments have shown their commitments by setting clear sustainability targets, offering tax incentives for green investments, and imposing regulations that encourage the adoption of sustainable practices (Salvador & Sancho, 2021). Further, governments are also very instrumental in supporting public funding and support for research and development in green technologies which are essential in overcoming initial financial and technological barriers (Wirba, 2023). Additionally, civil society organizations (CSOs) have also been key players in advocating for sustainable development. In most countries, these are very instrumental in conducting public awareness campaigns, and initiating grassroots initiatives, which translate to successful sustainable efforts (Martinez & Villa, 2023). Through their efforts, CSOs can influence policy, mobilize public opinion, advocate for evidence-based policy formulation, and foster community engagement in sustainability projects.

In their own right, each of the above organizations plays a key role in building a more informed and engaged public that supports and participates in sustainable practices. As for the scientific and academic community, they contribute significantly to the advancement of sustainable technologies through research, development, and dissemination of information and knowledge (Sarpong et al., 2023). Universities and research institutions serve as incubators for innovation, exploring new methods of energy production, waste reduction, and resource management. The dissemination of research findings and the promotion of knowledge exchange guarantees that advancements in sustainability reach a broad audience, including policymakers, industry leaders, and the general public, thereby accelerating the adoption of sustainable solutions (Sieg et al., 2023). For instance, it should be acknowledged that

Yale University, USA is at the forefront in offering sustainability-focused courses and programs, while the University of Cambridge, UK is credited with establishing the Centre for Circular Economy where research and collaborating with businesses to promote circular designs and waste reduction (University of Cambridge).

At the global scale, international organizations (such as the United Nations, the World Bank, African Development Bank) and collaborations play a vital role in facilitating knowledge exchange and coordinating global initiatives aimed at sustainability. Through platforms such as the United Nations Framework Convention on Climate Change (UNFCCC) and the Sustainable Development Goals (SDGs), international bodies provide a forum for sharing best practices, negotiating agreements, and mobilizing resources to address global environmental challenges. These collaborative efforts underscore the interconnectedness of sustainability issues and the need for a unified global response. As for Public–private partnerships, they exemplify successful collaborations that leverage the strengths and resources of both the public and private sectors to achieve sustainable outcomes. They particularly play a critical role in offering financial support to support public programs that governments would otherwise struggle or fail to implement (Fell & Mattsson, 2021). By combining government support with corporate expertise and investment, PPPs can undertake large-scale sustainability projects that might otherwise be unfeasible. Examples of such partnerships include urban renewal projects, renewable energy initiatives, and sustainable transportation systems, which have demonstrated significant environmental, economic, and social benefits (Bulkeley & Betsill, 2005).

6 Conclusion

The discussion in this chapter has delved into how technological innovation stands as the bedrock upon which sustainable transitions are built. It has been highlighted how innovations in renewable energy projects, smart grid technologies, and circular economy practices offer transformative solutions to the pressing environmental challenges confronting the world during this era. From the discussion, it has been established that these technologies not only provide the tools necessary to reduce carbon emissions and enhance energy efficiency but also empower societies to manage resources more sustainably, underscoring the pivotal role innovation plays in driving the shift towards sustainability. The synergies from these innovations facilitate a more resilient and flexible energy system, optimize resource use, and foster an economic model that emphasizes durability, reuse, and waste reduction. Such a comprehensive strategy not only accelerates the transition to sustainable urban environments but also ensures that these transitions are inclusive and equitable.

It should not be lost that the path to a sustainable future is not without notable challenges. In this chapter, challenges such as technological hurdles, financial constraints, regulatory barriers, and cultural resistance to change have been comprehensively discussed. In the course of the discussion, it has been established that overcoming these obstacles requires innovative solutions, strategic investments, supportive policy frameworks, and a paradigm shift in societal values towards sustainability. The need to address these challenges head-on has been emphasized, especially to facilitate the unlocking of the full potential of sustainable technologies and ensure their widespread adoption and implementation. Among solutions evaluated and discussed include the need for unprecedented collaboration across all sectors of society. The emphasis is that governments, businesses, civil society organizations, the scientific community, and international bodies must work together to create an enabling environment for sustainable technologies. This multi-sectoral approach fosters the exchange of knowledge, leverages financial and human resources, and ensures that sustainability initiatives are coordinated and aligned with global sustainability goals, thereby enhancing the effectiveness and impact of these efforts.

Going into the future, the landscape of sustainable technology is poised for rapid evolution, driven by continuous innovation and changing societal demands. Emerging trends, such as the digitization of energy systems, advances in material science, and the growth of green finance, promise to further revolutionize sustainable transitions. As such, staying abreast, or ahead of these trends requires ongoing research, adaptability, and a commitment to lifelong learning, exploration, and innovation within the sustainability sector. This is justified by the fact that the journey towards sustainability is an ongoing process that demands persistence, creativity, and a willingness to explore uncharted territories. Therefore, as humanity navigate this path, it is imperative to remain open to new ideas, embrace innovative approaches, and foster a culture of experimentation and learning. The quest for sustainable urban transitions is not just about adopting new technologies but also about reimagining the relationship with the environment and each other. It is a journey that calls for collective action, shared responsibility, and an unwavering commitment to building a more sustainable, resilient, and equitable world for future generations.

References

Allam, Z. (2019). Enhancing Renewable Energy Adoption in Megacities through Energy Diversification, Land Fragmentation and Fiscal Mechanisms. *Sustainable Cities Society, 53*(2), 101841.

Allam, Z. (2020a). *Surveying the Covid-19 Pandemic and Its Implications: Urban Health*. Elsevier.

Allam, Z. (2020b). *Urban Governance and Smart City Planning: Lessons from Singapore*. Emerald Group Publishing.

Allam, Z. (2020c). Vital COVID-19 Economic Stimulus Packages Pose a Challenge for Long-Term Environmental Sustainability. In *Surveying the Covid-19 Pandemic its Implications* (pp. 97–105). Elsevier.

Allam, Z. (2021). The Case for Autonomous Smart Cities in the Wake of Climate Change. In *The Rise of Autonomous Smart Cities* (pp. 61–74). Palgrave Macmillan.

Allam, Z., Chabaud, D., Gall, C., Pratlong, F., & Moreno, C., et al. (2023a). Chapter 5—Exploring the Relationship of Time Keeping and Urban Morphology within the Economic Renaissance and the Postmodern Era. In Z. Allam, D. Chabaud, & C. Gall (Eds.), *Resilient and Sustainable Cities* (pp. 77–91). Elsevier.

Allam, Z., Chabaud, D., Gall, C., Pratlong, F., & Moreno, C., et al. (2023b). Chapter 8—Financing the 15-minute City Concept and Its Infrastructural Ecosystem in Developing Nations Through Fiscal Mechanisms. In Z. Allam, D. Chabaud, & C. Gall (Eds.), *Resilient and Sustainable Cities* (pp. 121–133). Elsevier.

Allam, Z., Cheshmehzangi, A., & Jones, D. S. (2023c). Redefining Climate Policy and Economic Resilience in the Consuming World. *Climate and Social Justice: The Political Economy of Urban Resilience and Mercantilism* (pp. 77–102). Springer Nature.

Allam, Z., & Dhunny, Z. A. (2019). On Big Data, Artificial Intelligence and Smart Cities. *Cities, 89*, 80–91.

Allam, Z., Jones, D., & Thondoo, M. (2020). Economically Incentivizing Urban Sustainability and Resilience. In *Cities and Climate Change* (pp. 83–106). Palgrave Macmillam.

Allam, Z., & Newman P. (2023). Smart Cities and the Urban Economy: How Smart City Can Integrate Good Urban Economies. In *Revising Smart Cities with Regenerative Design* (pp. 101–116). Springer International Publishing.

Allam, Z., Sharifi, A., Bibri, S. E., & Chabaud, D. (2022a). Emerging Trends and Knowledge Structures of Smart Urban Governance. *Sustainability, 14*, 5275.

Allam, Z., Sharifi, A., Giurco, D., & Sharpe, S. A. (2021). On the Theoretical Conceptualisations, Knowledge Structures and Trends of Green New Deals. *Sustainability, 13*, 12529.

Allam, Z., Sharifi, A., Giurco, D., & Sharpe, S. A. (2022b). Green New Deals Could Be the Answer to COP26's Deep Decarbonisation Needs. *Sustainable Horizons, 1*, 100006.

Andreadakis, S., & Owusu-Wiredu, P. (2023). *Fashion Footprint: How Clothes Are Destroying Our Planet and the Growing Impacts of Fast Fashion*. IntechOpen.

Austin Energy. (2019). *Integrated Smart Grid: Reinventing Electric Service Delivery with Innovative Technologies*. Available at: https://austinenergy.com/about/company-profile/electric-system/integrated-smart-grid.

Awan, U., & Sroufe, R. (2022). Sustainability in the Circular Economy: Insights and Dynamics of Designing Circular Business Models. *Applied Sciences, 12*, 1521.

Barragán-Escandón, A., Terrados-Cepeda, J., & Zalamea-León, E. (2017). The Role of Renewable Energy in the Promotion of Circular Urban Metabolism. *Sustainability, 9*, 2341.

Bulkeley, H., & Betsill, M. (2005). Rethinking Sustainable Cities: Multilevel Governance and the 'Urban' Politics of Climate Change. *Environmental Politics, 14*, 42–63.

Burundtland, G. H. (1987). *Our Common Future: Report of the World Commission on Environment and Development*. Oxford University.

Castka, P. (2020). The Role of Standards in the Development and Delivery of Sustainable Products: A Research Framework. *Sustainability, 12*, 10461.

Catalbas, M. C. (2022). Impacts of COVID-19 Pandemic on Electrical Energy Storage Technologies. *Energy Storage, 4*, e305.

Chabaud, D., Allam, Z., Gall, C., Pratlong, F., & Moreno, C. (2023). Redefining Investable Infrastructure in Developing Nations in a Postpandemic Era: The Case of the 15-Minute City. In *Resilient and Sustainable Cities* (pp. 136–149). Elsevier.

Cheshmehzangi, A. (2020). The Analysis of Global Warming Patterns from 1970s to 2010s. *Atmospheric and Climate Sciences*, 10(3). https://doi.org/10.4236/acs.2020.103022

Cheshmehzangi, A., & Dawodu, A. (2019). *Sustainable Urban Development in the Age of Climate Change - People: The Cure or Curse*. Palgrave Macmillan.

Cheshmehzangi, A., Dawodu, A., & Sharifi, A. (2021). *Sustainable Urbanism in China*. Routledge.

Cheshmehzangi, A., You, N., Siri, J., & Birch, E. (2024). *Harnessing Urban Innovation to Unlock the Sustainable Development Goals*. Springer.

Choi, D., Chung, C. Y., Seyha, T., & Young, J. (2020). Factors Affecting Organizations' Resistance to the Adoption of Blockchain Technology in Supply Networks. *Sustainability, 12*, 8882.

de Reuver, M., van der Lei, T., & Lukszo, Z. (2016). How Should Grid Operators Govern Smart Grid Innovation Projects? An Embedded Case Study Approach. *Energy Policy, 97*, 628–635.

Dhunny, A., Doorga, J., Allam, Z., Lollchund, M. R., & Boojhawon, R. (2019). Identification of Optimal Wind, Solar and Hybrid Wind-solar Farming Sites Using Fuzzy Logic Modelling. *Energy, 188*, 116056.

Ding, S., Tukker, A., & Ward, H. (2023). Opportunities and Risks of Internet of Things (IoT) Technologies for Circular Business Models: A Literature Review. *Journal of Environmental Management, 336*, 117662.

Elabbassi, I., Khala, M., El yanboiy, N., Eloutassi, O., & Hassouani, Y. El. (2024). Evaluating and Comparing Machine Learning Approaches for Effective Decision Making in Renewable Microgrid Systems. *Results in Engineering, 21*, 101888.

Ellen Macarthur Foundation. (2016). *The New Plastic Economy: Rethinking the Future of Plastics* (p. 120). Ellen Macarthur Foundation.

Ellen Macarthur Foundation. (2023). *The Circular Economy*. Available at: https://www.ellenmacarthurfoundation.org/the-circular-economy-in-detail-deep-dive#:~:text=A%20circular%20economy%20is%20a,the%20consumption%20of%20finite%20resources.

Fairphone. (2024). *Longevity*. Available at: https://www.fairphone.com/en/impact/long-lasting-design/.

Fehrer, J. A., Kemper, J. A., & Baker, J. J. (2024). Shaping Circular Service Ecosystems. *Journal of Service Research, 27*, 49–68.

Fell, T., & Mattsson, J. (2021). The Role of Public-Private Partnerships in Housing as a Potential Contributor to Sustainable Cities and Communities: A Systematic Review. *Sustainability, 13*, 7783.

Gupta, R., & Chaturvedi, K. T. (2023). Adaptive Energy Management of Big Data Analytics in Smart Grids. *Energies, 16*, 6016.

Hoehn, D., Laso, J., Margallo, M., et al. (2021). Introducing a Degrowth Approach to the Circular Economy Policies of Food Production, and Food Loss and Waste Management: Towards a Circular Bioeconomy. *Sustainability, 13*, 3379.

IEA. (2023a). *Tracking Smart Grids*. Available at: https://www.iea.org/energy-system/electricity/smart-grids.

IEA. (2023b). *Will Solar PV and Wind Costs Finally Begin to Fall Again in 2023 and 2024?* Available at: https://www.iea.org/reports/renewable-energy-market-update-june-2023/will-solar-pv-and-wind-costs-finally-begin-to-fall-again-in-2023-and-2024.

IEA. (2023c). *World Energy Employment*. Available at: https://www.iea.org/reports/world-energy-employment-2023/executive-summary.

International Renewable Energy Agency. (2020). *Renewable Energy and Jobs—Annual Review 2020*. IRENA.

IPCC. (2023). Sections. In Core Writing Team, H. Lee, & J. Romero (Eds.), *Climate Change 2023: Synthesis Report. Contribution of Working Groups I, II and III to the Sixth Assessment Report of the Intergovernmental Panel on Climate Change* (pp. 35–115). IPCC.

Jordan, A., & Lenschow, A. (2010). Environmental Policy Integration: A State of the Art Review. *Environmental Policy and Governance, 20*, 147–158.

Kabeyi, M. J. B., & Olanrewaju, O. A. (2023). Smart Grid Technologies and Application in the Sustainable Energy Transition: A Review. *International Journal of Sustainable Energy, 42*, 685–758.

Kirmani, S., Mazid, A., Khan, I. A., & Abid, M. (2023). A Survey on IoT-Enabled Smart Grids: Technologies, Architectures, Applications, and Challenges. *Sustainability, 15*, 717.

Lazer, L., Khandelwal, N., & Wellman, J. (2020). Why is Sustainable Urban Transport a Great Investment? *World Bank Blog*. The World Bank.

Levis Strauss & Co. (2019). *How Levi's is Saving Water*. Available at: https://www.levistrauss.com/2019/03/25/world-water-day-2019-saving-h2o/.

Lipp, J. (2007). Lessons for Effective Renewable Electricity Policy from Denmark, Germany and the United Kingdom. *Energy Policy, 35*, 5481–5495.

Mahmoud, M. A., Md Nasir, N. R., Gurunathan, M., Raj, P., & Mostafa, S. A. (2021). The Current State of the Art in Research on Predictive Maintenance in Smart Grid Distribution Network: Fault's Types, Causes, and Prediction Methods—A Systematic Review. *Energies, 14*(16), 5078.

Martínez-Peláez, R., Ochoa-Brust, A., Rivera, S., Félix, V. G., Ostos, R., Brito, H., Félix, R. A., & Mena, L. J. (2023). Role of Digital Transformation for Achieving Sustainability: Mediated Role of Stakeholders, Key Capabilities, and Technology. *Sustainability, 15*, 11221.

Martinez, A., & Villa, S. (2023). *Promoting Sustainable Development through Civil Society*. IMF International Monetary Fund.

Moreno Escobar, J. J., Morales Matamoros, O., Tejeida Padilla, R., Lina Reyes, I., & Quintana Espinosa, H. (2021). A Comprehensive Review on Smart Grids: Challenges and Opportunities. *Sensors (Basel), 21*, 6978.

Mostafa, N., Ramadan, H. S. M., & Elfarouk, O. (2022). Renewable Energy Management in Smart Grids by Using Big Data Analytics and Machine Learning. *Machine Learning with Applications, 9*, 100363.

Nadeem, T. B., Siddiqui, M., Khalid, M., & Asif, M. (2023). Distributed Energy Systems: A Review of Classification, Technologies, Applications, and Policies. *Energy Strategy Reviews, 48*, 101096.

Oraee, A., Pohl, L., Geurts, D., & Reichel, M. (2024). Overcoming Premature Smartphone Obsolescence amongst Young Adults. *Cleaner and Responsible Consumption, 12*, 100174.

Pathan, M. S., Richardson, E., Galvan, E., & Mooney, P. (2023). The Role of Artificial Intelligence within Circular Economy Activities—A View from Ireland. *Sustainability, 15*, 9451.

Pauw, J.B.-d, Gericke, N., Olsson, D., & Berglund, T. (2015). The Effectiveness of Education for Sustainable Development. *Sustainability, 7*, 15693–15717.

Pollitt, M. (2012). The Role of Policy in Energy Transitions: Lessons from the Energy Liberalisation Era. *Energy Policy, 50*, 128–137.

Renugadevi, N., Saravanan, S., & Naga Sudha, C. M. (2023). IoT Based Smart Energy Grid for Sustainable Cites. *Materials Today: Proceedings, 81*, 98–104.

Sabbati, G. (2021). *At a Glance: Living in the EU-Circular Economy*. Globalstat.

Salet, W., & de Vries, J. (2019). Contextualisation of Policy and Law in Sustainable Urban Development. *Journal of Environmental Planning and Management, 62*, 189–204.

Salvador, M., & Sancho, D. (2021). The Role of Local Government in the Drive for Sustainable Development Public Policies. An Analytical Framework Based on Institutional Capacities. *Sustainability, 13*, 5978.

Sarpong, D., Boakye, D., Ofosu, G., & Botchie, D. (2023). The Three Pointers of Research and Development (R&D) for Growth-Boosting Sustainable Innovation System. *Technovation, 122*, 102581.

Shourkaei, M. M., Taylor, K. M., & Dyck, B. (2024). Examining Sustainable Supply Chain Management Via a Social-symbolic Work Lens: Lessons from Patagonia. *Business Strategy and the Environment, 33*, 1477–1496.

Sieg, P., Posadzińska, I., & Jóźwiak, M. (2023). Academic Entrepreneurship as a Source of Innovation for Sustainable Development. *Technological Forecasting and Social Change, 194*, 122695.

Sjödin, D., Parida, V., & Kohtamäki, M. (2023). Artificial Intelligence Enabling Circular Business Model Innovation in Digital Servitization: Conceptualizing Dynamic Capabilities, AI Capacities, Business Models and Effects. *Technological Forecasting and Social Change, 197*, 122903.

Strielkowski, W., Civín, L., Tarkhanova, E., Tvaronaviciene, M., & Petrenko, Y. (2021). Renewable Energy in the Sustainable Development of Electrical Power Sector: A Review. *Energies, 14*, 8240.

Sylla, S., & Springmann, E. (2023). *The Smart Meter Rollout in Germany and Europe*. Available at: https://www.ffe.de/en/publications/the-smart-meter-rollout-in-germany-and-europe/.

The World Bank. (2023). From Evolution to Revolution: 10 Years of Green Bond Event. *Green Bonds*. The World Bank.

UNCTAD. (2023). Data for Development. *United Nations Commission on Science and Technology for Development* (pp. 1–82). UNCTAD.

Unilever. (2021). Unilever Sustainable Living Plan 2010 to 2020: Summary of 10 year's progress. Unilever, 1–39.

United Nations. (2020). *United Nations Sustainable Development Goals*. Available at: https://sdgs.un.org/goals.

University of Cambridge. *Circular Economy Centre*. Available at: https://www.jbs.cam.ac.uk/faculty-research/centres/circular-economy/.

Whitehead, D., Owens, K., Gammel, D., & Smith, J. (2017) *Ukraine Cyber-Induced Power Outage: Analysis and Practical Mitigation Strategies*. IEEE.

Wirba, A. V. (2023). Corporate Social Responsibility (CSR): The Role of Government in promoting CSR. *Journal of the Knowledge Economy*.

Yamoah, F. A., Sivarajah, U., Mahroof, K., & Peña, I. G. (2022). Demystifying Corporate Inertia Towards Transition to Circular Economy: A Management Frame of Reference. *International Journal of Production Economics, 244*, 108388.

CHAPTER 4

The Social Movements and Political Landscape of Sustainable Futures

Abstract Social movements and political dynamics profoundly shape the pathway towards sustainable futures. This chapter delves into the interplay between grassroots activism and political realities, examining how these forces drive societal transitions towards sustainability. Social movements, defined as organized efforts by large groups to achieve social change, and grassroots activism, characterized by community-driven initiatives, play pivotal roles in this context. Recent global challenges, such as climate change and the COVID-19 pandemic, underscore the urgency of this discourse. The interconnectedness between social movements and policymaking is critical for realizing sustainable futures. Movements like the Green New Deal (GND) exemplify how grassroots activism can influence political landscapes and promote sustainable urban transitions. Social movements not only advocate for environmental justice but also empower communities through direct involvement in decision-making processes. This chapter also explores the strategies employed by social movements, including direct actions, advocacy campaigns, and community mobilization. The impact of these movements is illustrated through case studies like the Fridays for Future campaign and the Right to Repair movement, showcasing their influence on policy and public opinion. Despite their successes, social movements face challenges such as resource constraints, political resistance, and institutional inertia. Addressing these obstacles requires innovative approaches, coalition-building, and leveraging digital platforms for global collaboration.

Z. Allam and A. Cheshmehzangi, *Sustainable Futures and Green New Deals*, https://doi.org/10.1007/978-3-031-63642-4_4

Keywords Social Movements · Grassroots Activism · Sustainable Development · Political Landscape · Climate Change · Environmental Justice · Community Mobilization · Policy Influence

1 Introduction

The pathway towards sustainable futures winds through intricate landscapes influenced by the dynamics interplay between social movements (both grassroots and civil societies) and political realities. This chapter will explore these, not just as an examination of a merely academic exercise, but with the intention to understand the very fabric of how societies can transition to more sustainable modes of living. The core of this discourse are key concepts such as social movements, grassroots activism, and sustainable development.

Social movements are defined in a variety of ways, such as organized efforts by a large group of people to achieve a common goal, often focused on social change (Diani, 1992). Grassroots activism on the other hand refers to action on social issues that are initiated at a local or community level, driven by the collective participation of ordinary citizens (Cress & Snow, 2000). On its part, sustainable development is defined by the Brundtland Commission as '*development that meets the needs of the present without compromising the ability of future generations to meet their own needs*' (Brundtland, 1987). With recent progressions in sustainable development and innovations (Cheshmehzangi et al., 2024), we also believe equipping the future generations (Cheshmehzangi & Dawodu, 2019a; Cheshmehzangi et al., 2021) would help to find new pathways and social movements to change the political landscape of sustainable futures.

Following a series of catastrophes and pandemics in the recent past particularly prompted by climate change and COVID-19 (Allam, 2020a, 2020b, 2020c), the urgency of this discourse cannot be overstated. The seriousness of the situation is well defined in a report by the Intergovernmental Panel on Climate Change (IPCC, 2023) highlighting the critical need for immediate and unprecedented changes to all aspects of society to limit global warming to 1.5 C above pre-industrial levels (Cheshmehzangi, 2020; Cheshmehzangi & Chen, 2021). While the report acknowledges that this target might be a toll order, following a myriad of issues, temperatures beyond this limit might lead to unprecedented outcomes. This underscores the urgency of addressing environmental challenges through collective action of all stakeholders. The emphasis on urgency is also pronounced in the United Nations' Sustainable Development Goals (SDGs) which underscores the interconnectedness of social, economic, and environmental sustainability, thus reinforcing the need for a holistic approach to sustainable development (United Nations, 2020).

Guided by the above reports, it is worth noting that the interconnectedness between social movements and policymaking processes is a critical area of focus for sustainable futures to be realized. Social movements have the potential to influence both the implementation and enforcement of policies, including the ones mentioned above by raising awareness, shaping public opinion, and directly engaging with policymakers (Buchter, 2023). This influence is evident in the transition towards more sustainable urban futures, where grassroots activism plays a crucial role in promoting policies that support sustainable development, renewable energy adoption, and the equitable distribution of resources (Skaidrė, 2019).

The emergence of the Green New Deal (GND) serves as a contemporary example of how social movements and political landscapes intersect with the pursuit of sustainable urban transitions. Originating in the United States, after a spirited agitation by social movement and eventually being embraced by the political class, and finally gaining traction globally, the GND outlines a plan for transitioning to 100% renewable energy within a decade while ensuring social justice and equity through job creation and infrastructure investment (Rifkin, 2019; Allam et al., 2021, 2022b). It should be understood that the GND is not just a policy proposal; it is a manifestation of years of grassroots activism and social movements demanding action on climate change and social inequality (Calhoun & Fong, 2022).

At the community level, local movements, not only contribute to the reduction of carbon emissions but also empower communities by involving them directly in the decision-making process (Cheshmehzangi & Tang, 2022). This empowerment is crucial for sustainable development, as it ensures that the transition to sustainable practices such as the adoption of renewable energy is both equitable and inclusive.

The role of social movements in sustainable urban transitions is also evident in their push for urban regeneration projects that prioritize sustainability, social justice, and economic equity (Allam et al., 2023f).

A case in point is the role social movements played, by pressurizing the government to revitalize the Cheonggyecheon stream, under the Cheonggyecheon Stream Restoration Project. As a result, the city is now safe and free from air pollution and residents are able to enjoy improved public health (Kim & Yang, 2023). In India, the 'Save Aarey Movement' is credited for successfully advocating against deforestation in the critical urban forests, thus contributing to the preservation of biodiversity and mitigating flood risks in the cities nearby (Dsouza et al., 2021).

Urban regeneration, when guided by the principles of sustainability, not only translates to revitalization of urban areas but also ensures that such an exercise benefits all members of the community and minimizes environmental impact (Allam & Newman, 2018; Allam, 2019; Allam & Raisah Takun, 2022; Bosone et al., 2023). A sizeable number of public projects that incorporate green spaces, sustainable transportation, and energy-efficient buildings are often the result of advocacy by social movements that recognize the importance of sustainable urban development. For instance, a case can be made that the '15-Minute City' movement successfully convinced Mayor Anne Hidalgo of Paris, France to implement the concept in the city (Allam et al., 2022a, 2023b, 2023d), leading to the city being among pioneers in sustainable urban transition with reduction of motor vehicles in the city by almost 50% (Abdelfattah et al., 2022; Moreno et al., 2021). The influence of social movements on sustainable urban transitions (Cheshmehzangi & Dawodu, 2019b, 2019c) can also be seen in the promotion of circular economy principles. For instance, the *Right to Repair Movement* which advocates the right of consumers to repair their own electronic devices and access spare parts; hence, challenging the planned obsolescence model is a notable example to note (Tischner & Stasiuk, 2023). The circular economy emphasizes the reduction, reuse, and recycling of materials, and offers a pathway to sustainable development by minimizing waste and conserving resources (Awan & Sroufe, 2022). Social movements advocating for zero waste and sustainable consumption practices play a critical role in pushing for the adoption of circular economy principles in urban areas as will further be highlighted in the successive sections below.

2 Social Movements for Sustainable Futures

The scope of social movements advocating for sustainable futures is both vast and dynamic, covering a wide array of initiatives and campaigns worldwide. Those include youth activism, indigenous resistance, community land trusts, transition towns movement and climate justice movements that stand out as powerful forces driving the environmental, and social change agenda at the global level.

In youth activism, there are numerous notable examples that have risen in the recent past. Among these is the global *Fridays for Future* campaign started by Greta Thunberg. Since it started, the movement has mobilized millions of young people across continents to demand urgent action to combat climate change and advocate for sustainable development practices (Fridays for Future, 2018). Another notable example is the *Sunrise Movement* based in the United States, which advocates for GND and holds politicians accountable for climate action (Sunrise Movement, 2023). There is also the *Ayana Elizabeth Johnson movement* led by Ayana herself to advocate for climate justice and ocean conservation, particularly in view of the disproportionate impact climate change has had on vulnerable communities (Mccarthy, 2021).

Indigenous resistance movements, which too have a long and cultivating history are also diverse, and encompass a wide range of groups from different parts of the world fighting for their land, rights, and cultural survival. In most cases, these movements are associated with struggles for sustainability, more so focusing on protecting their traditional territories and ancestral heritage and knowledge. Examples of such groups include the *Standing Rock Sioux Tribe*, which is opposed to the Dakota Access Pipeline in the United States and highlights the fight for clean water and opposition to fossil fuel projects on indigenous lands (National Museum of the American Indian, 2018). The Sami people in Scandinavia and their struggle for their cultural rights like reindeer herding practices and land ownership is another notable example (Heaps, 2019). In South America, there are the Amazonian tribes' movements, with diverse indigenous groups from the Amazon Basin that fight against deforestation, extractive industries, and non-inclusive government policies that not only threaten their habitats and territories but also their livelihoods (Regional Health-Americas, 2022). These examples underscore the pivotal role indigenous communities play in protecting natural resources and preserving their rights against exploitation (Whyte, 2018). Climate justice movements, which seek to address the disproportionate impact of climate change on marginalized communities, have gained prominence, advocating for equitable solutions to environmental issues (Schlosberg & Collins, 2014).

2.1 Strategies Employed by Movements

The above-mentioned movements leverage a variety of strategies to achieve their objectives, have their voices heeded, and recruit even more members. Such strategies include direct actions, spirited advocacy campaigns, and community mobilization and organizing. Direct actions take different forms including public demonstrations, non-violent protests, sit-ins and occupations, and boycotts and strikes among others serve as a critical tool for drawing attention to pressing environmental issues and exerting pressure on policymakers and corporations (della Porta & Diani, 2006).

Advocacy campaigns (e.g. the Greta Thunberg's school strikes, Fracking bans advocacy in different regions and others), often orchestrated in collaboration with non-governmental organizations (NGOs), focus on influencing policy development and promoting sustainable practices through lobbying, public relations efforts, and educational programs (Hadden, 2015). On the other hand, community organizing involves mobilizing local communities to advocate for changes in local policies and practices, emphasizing grassroots participation and empowerment (Christens et al., 2021). Hence, possibilities for further social movements and bringing people back to the decision-making table could help develop better public relations but just pseudo participation.

3 Impact of Social Movements: Case Studies

The impacts of these movements on environmental and social policy have been growing significantly in recent years. For example, the divestment movement, spearheaded by 350.org, has successfully persuaded over 1,110 institutions to commit to divesting $11 trillion from fossil fuels, highlighting the economic leverage of collective action (Tyler-Davies, 2019). Another example is the urban agriculture movement such as 'Growth in Detroit', which has transformed vacant lots into productive gardens in cities like Detroit, improving local food security and fostering community resilience (Newell et al., 2022). Further, the *Right to Repair*

Movement highlighted in the previous section that advocates for the right of consumers to repair their own electronic devices has managed to force the political class in several U.S. states and the European Union to consider formulating relevant policies, which would be crucial in the sustainability agenda (Tischner & Stasiuk, 2023).

3.1 *Diversity within Social Movements*

Social movements are known to draw their strength from their inclusivity and diversity. Integrating the voices and perspectives of marginalized communities, including indigenous peoples, people of colour, and low-income populations allows these movements to enrich their discourse on sustainability and ensure that solutions are equitable and just. This inclusivity not only broadens the base of support for environmental initiatives but also addresses intersectional issues of environmental justice. This is critical, particularly recognizing that social inequality and environmental degradation are deeply interconnected (Newell et al., 2022).

3.2 *Role of Social Media and Digital Platforms*

The fourth industrial revolution, characterized by advanced technology development, prompted the rise and infiltration of diverse social media platforms. Through these plus digital platforms, social movements have crafted ingenious methodologies and approaches for them to organize, communicate, and mobilize. Social media platforms have enabled activists to reach a global audience, share information rapidly, and coordinate actions across geographical boundaries devoid of traditional barriers and constraints (Eren-Erdoğmuş & Ergun, 2017). The use of social media in movements like Fridays for Future has amplified its message, attracted widespread media coverage, and engaged a broader demographic in environmental activism.

4 Challenges and Opportunities

4.1 *Challenges Hindering the Effectiveness of Social Movements*

Social movements have been gaining popularity and achieving groundbreaking outcomes despite numerous challenges. Internally, social movements have been hampered by resource constraints, such as funding and staffing which limits their ability to sustain long-term campaigns and achieve their objectives. They are also prone to internal divisions, often arising from differing priorities and strategies, leadership wrangles, and unnecessary competitions within diverse coalitions hence, impeding their collective action and weakening movement cohesion (Larmer, 2010). A practical case can be built to affirm this by exploring the unfortunate decline of the *Occupy Wall Street* Movement, which was agitating for economic equity, but was marred by a lack of diverse ideologies and a lack of clear leadership structure (Roberts, 2012). Furthermore, resistance from powerful interests, including government, corporate entities, and influential individuals poses significant obstacles to achieving policy change and implementing sustainable practices (Jasper, 2018).

Externally, a major setback to the success of social movements is the lack of political goodwill. This can be pointed out by several factors like the reluctance by the policymakers to enact necessary environmental regulations, possibly stemming from economic apprehensions or political pressures. This reluctance is often influenced by short-term economic considerations at the expense of long-term sustainability goals, reflecting a systemic preference for immediate gains over future well-being (Woo & Kang, 2020).

Institutional inertia is another hindrance that exacerbates the woes of social movements, exhibiting resistance within organizations and governmental bodies to alter entrenched operating procedures and norms (Barnett et al., 2015; Beunen & Patterson, 2019). This inertia is not merely a product of bureaucratic sluggishness but arises from selfish interests that benefit from maintaining the status quo, creating a formidable obstacle to the adoption of innovative sustainability practices.

The backlash from vested interests represents another significant challenge. Industries that perceive sustainability transitions as a threat to their economic interests, notably fossil fuel companies and transport-based companies employ lobbying and disinformation campaigns to undermine environmental initiatives. For instance, in late 2023, the EU proposed to increase the ambition of its Emission Trading Systems (ETS), mostly targeting sectors like transport and buildings. However, such efforts were strongly opposed by the airline industry lobbying against the inclusion of the aviation sector in the ETS, citing issues like increased operation costs and competitiveness (Schvartzman, 2023). These campaigns not only distort public discourse but also influence policy decisions, creating an environment where misinformation impedes progress toward sustainability.

By addressing these challenges with strategic, innovative, and inclusive approaches, social movements can navigate the complex landscape of sustainability advocacy, gradually overcoming obstacles to effect meaningful environmental and social change. The synergy of collective action, empowered by education and innovation, holds the potential to drive the transformative change necessary for a sustainable future.

4.2 *Opportunities for Overcoming Challenges*

The daunting obstacles discussed above do not necessarily weather down the social movements' avenues for advancing their agendas. In most cases, they find refuge in coalition-building and the formation of strategic alliances, which enables them to amplify their voices and consolidate resources. By uniting diverse groups with a common interest in sustainability, these coalitions can exert greater influence on policy and public opinion, overcoming the fragmentation that often weakens individual efforts (Van Dyke & McCammon, 2010).

Another potent avenue that has been exploited by social movements is the collective action approach empowered by global communication networks such as social media platforms. This interconnectedness enables the synchronization of initiatives across national borders, fostering a global response to what are inherently global challenges (Eren-Erdoğmuş & Ergun, 2017). Such unity is critical in confronting multinational corporations and influencing international policy, demonstrating the power of collective global action in the face of localized institutional barriers.

4.3 *Role of Intersectionality*

The concept of intersectionality, which can be understood as the interconnectedness of diverse social identities (social movements in this paper) and their combined impacts on individuals or communities is pivotal in ensuring that the movement towards sustainable futures is inclusive and equitable. Recognizing the complex interplay of race, class, gender, and other social categorizations allows social movements to address the deeper systemic causes of environmental degradation and social injustice. This holistic approach not only broadens the appeal of sustainability movements but also ensures that the solutions proposed do not inadvertently perpetuate existing disparities but rather contribute to a more equitable society (Roth, 2021; Hiraide and Evans). Hence, the importance of social justice (Allam et al., 2023g) represents how social movements and diverse social identities are tangled together.

5 Innovative Approaches and Best Practices

In view of the diverse challenges and barriers that social movements face, they cannot slack in exploring ways to remain strong and focused. Innovation in engagement and mobilization tactics is among the potent pathways they can pursue for the success of sustainability initiatives. Innovations could be deployed to facilitate participatory decision-making processes such as citizen assemblies and community-led planning, democratizing the discourse on sustainability, and ensuring that projects reflect the diverse needs and values of the community (Honeybun-Arnolda et al., 2024). Similarly, they can leverage digital platforms for knowledge sharing and collaboration allowing for the rapid dissemination of best practices and innovative solutions, facilitating a dynamic exchange of ideas that can be tailored to local contexts.

Another approach that they can adopt is the adoption of arts and storytelling strategies, which emerge as powerful tools for conveying the urgency of the sustainability crisis. Social movements can exploit these to engage emotions and imagination; hence inspiring action and commitment among a broad audience, transcending the limitations of traditional advocacy and education methods.

5.1 Decentralized, Bottom-up Approaches

The shift towards decentralized, bottom-up approaches marks a significant evolution in the strategy for achieving sustainable futures. This is another strategy that social movements need to deploy, especially in their quest to garner maximum local support and legitimacy. Emphasizing local knowledge and leadership would not only benefit the social movements but also have the potential to empower communities to craft solutions that are uniquely suited to their environmental and social contexts. The locally inspired solutions not only enhance the resilience and sustainability of proposed projects and programs but also ensure that they are equitable, reflecting the needs and aspirations of the community (Kaiser, 2020). Hence, participatory processes help create transitional transformations that put people at the heart of achieving sustainable development and sustainable futures.

5.2 *Role of Education and Awareness-Raising*

Social movements benefit society through education and awareness-raising, which serve as the foundation for empowering individuals to engage with sustainability issues. They manage to do this through comprehensive education initiatives, media campaigns, and public outreach, targeting to equip individuals and the community at large with the knowledge and skills necessary for informed environmental stewardship. This empowerment is crucial for cultivating a broad-based movement capable of advocating for systemic change, both locally and globally. A case can be drawn from the Kokrobite Beach, Ghana Clean-up Initiative which entails regular clean-up events that involve youth and community members. The initiative is largely influenced by educational workshops, creative campaigns, and collaboration with local institutions such as schools (Fosu-Mensah & Yirenya-Tawiah, 2014). Such collaboration would create a sort of bottom-up capacity-building beyond just raising public awareness, and help creating continuing in educating society at a faster pace.

5.3 *Influencing Policy and Promoting Change*

The discussion above has comprehensively showcased that social movements leverage various mechanisms to impact policymaking processes. Over the years, the most successful mechanism has been direct lobbying, which involves a targeted effort to influence policymakers. The basis for this approach involves presenting research, drafting policy proposals, and directly engaging in discussions with legislators to persuade them of the movement's perspective (Berry, 2015; Yoshioka, 2010). This approach is different from grassroots mobilization, which relies on the power of collective action at the local level, where organized efforts lead to widespread public support. Grassroots mobilizations target to exert pressure on policymakers through petitions, town hall meetings, and other forms of civic engagement, highlighting the public's demand for change (Vasoo, 1991). Public pressure, significantly amplified by media coverage and large-scale demonstrations, boycotts, and other outcomes, places policymakers in a position where ignoring the movement's demands could result in compromising their political careers.

From a historical background, the social movements could borrow some tips from the civil rights movements which have had substantial success in advocacy utilizing grassroots mobilization, direct action, and strategic litigation to combat racial segregation and discrimination. This multifaceted approach has led to landmark policy reforms in different epochs of global history. Examples include the Civil Rights Act of 1964 (Act, 1964) and the Voting Rights Act of 1965 (Davidson, 1992) in the United States which fundamentally altered the legal landscape in favour of civil rights.

Another example that could influence the social movements in their pursuit is the practices of the environmental movements which also use strategies such as direct lobbying, public demonstrations, and legal challenges to prompt significant environmental legislation in different parts of the world. For instance, it is the efforts of such movements that led to the establishment of the EPA and the enactment of the Clean Air and Clean Water Acts (Daniels et al., 2019).

The above examples underline the potential of well-orchestrated advocacy efforts to achieve substantial policy changes. However, it should be noted that relying solely on top-down policy approaches presents several challenges that cannot be ignored. For instance, such approaches might not fully address the complexities of issues or the specific needs of affected communities. In most cases, they would lead to solutions that are either ineffective or may exacerbate social injustices, or lack sustainability (Rahman et al., 2023). This then justifies the need for grassroots activism (drawn from a bottom-up approach), which when grounded in the lived experiences of those directly impacted offers a more nuanced understanding of the issues at hand. However, these too need to be supported with sufficient resources and political goodwill, to ensure they do not struggle to effect broad policy change. In summary, this discussion highlights the need for a strategic blend of grassroots and formal political engagement in the sustainable future pursuit, as each has undeniable advantages and strengths.

Besides the adoption of multifaceted approaches, social movements need to be keen on policy feedback loops as these play a fundamental role in the dynamics and relationships between social movements and political and social changes (Schmid et al., 2020). Successful policy reforms can validate the efforts of social movements, encouraging increased activism and engagement. However, policy failures or setbacks may lead to demobilization or necessitate a strategic reassessment. It thus behoves the need to effectively leverage policy feedback loops to allow movements to adapt and maintain momentum, facilitating ongoing influence over policy and societal norms. Consulting with intermediaries such as NGOs and think tanks is also pivotal for social movements, particularly in connecting social movements with policymakers. These entities are rich in expertise, resources, and the necessary legitimacy to elevate grassroots demands to actionable policy considerations (Mosse & Nagappan, 2021). By translating complex issues into policy proposals and facilitating dialogue, intermediaries ensure that reforms are well-informed and encompass a wide array of perspectives.

Enhancing the effectiveness and sustainability of advocacy efforts is essential for long-term impact, particularly in initiatives such as sustainable futures which are greatly needed, especially by vulnerable communities and economies. It is then critical for entities such as social movements to build coalitions as such would enable them to combine resources and expertise, presenting a unified front that holds greater sway over policy discussions (Van Dyke & McCammon, 2010). Similarly, it would be advisable for them to adopt strategic framing which would help in communicating the movement's goals in a way that resonates with both the public and policymakers, making the cause more relatable and compelling (Caiani, 2023). Subsequently, long-term engagement in policy processes, including diligent monitoring and evaluation would also be important as such ensures that reforms are implemented as intended and adjusted according to evolving needs, securing sustained advancements toward the movement's objectives.

5.4 Political Actors and Interest Groups

The political actors, just like the social movements play a very critical role in the actualization of the journey towards sustainable futures. The complex interplay of various political actors and interest groups, each with its distinct agenda and degree of influence is what has shaped numerous groundbreaking initiatives, policies, and outcomes that guide different facets of the world even today. In the discourse on sustainability, diverse stakeholders such as government agencies (under the umbrella of the political actors), corporations, non-governmental organizations (NGOs), and grassroots organizations play critical roles in either promoting or obstructing sustainability initiatives. The dynamics between these entities, characterized by their power struggles, influence strategies, and collaborative efforts, underscore the challenges and opportunities in achieving sustainable development goals, more so in urban areas that are currently home to over 55% of the global population (Sun et al., 2020).

Government agencies are always at the forefront of formulating and enforcing sustainability policies, while at the same time, they are tasked with the delicate responsibility of ensuring a sound balance between economic growth and environmental protection. Their effectiveness, however, is double-pronged, in that it may be positively influenced or compromised by political pressures or powerful lobbying groups. On a positive note, when groups such as social movements and grassroot groups are in support, the government is able to achieve substantial projects that benefit the public. However, government agencies may be negatively influenced by interested groups to sway policy outcomes away from public interest towards more partisan or corporate interests; hence, jeopardizing the sustainability agenda, social justice, and economic welfare. In most cases, the prevailing political ideology significantly influences the agenda of government agencies, which can either facilitate the swift implementation of sustainability initiatives or create substantial barriers to their enactment. A good example is the unfortunate turnaround by the United States regarding its commitment to the Paris Agreement, which it eventually withdrew in 2020 (UNFCCC, 2020). However, with the change of government, the United States later (on 20 January 2021) resumed its original position on the accord (Government of United States, 2021).

As for the corporate sector, especially companies in the energy, agriculture, and manufacturing industries, the government wields significant influence over environmental policies. While a segment of the corporate world has progressively adopted sustainable practices and green technologies, a considerable portion continues to resist regulatory changes that would impose additional operational costs or necessitate significant alterations to their business models (Zhu et al., 2023). On this, it is understandable that corporations exercise their influence through lobbying efforts, political campaign contributions, and engagement in public–private partnerships, often aiming to shape policy outcomes in a manner that aligns with their business interests, sometimes at the expense of broader environmental and social objectives. When such are in close relationship with the government of the day, probably by having contributed to its election, they could wield significant power to resist any government-proposed policy, whether targeting environmental sustainability provided such may deem to disadvantage their business model.

In contrast, NGOs and grassroots organizations champion environmental protection and sustainable development from a standpoint of moral and ethical concern. These entities typically employ public campaigns, advocacy, and legal actions to push their agendas forward. Their strategies encompass mobilizing public opinion, direct engagement with policymakers, and the formation of coalitions to enhance their advocacy strength (Berry, 2015). Despite often operating with limited resources, these organizations are pivotal in raising awareness and advocating for change, contributing significantly to the global sustainability movement at both local and international levels, and pushing the enforcement of policies by the government without favouritism or bias.

Within government agencies, the influence of money in political processes, particularly through lobbying by affluent interest groups, plays a critical role in shaping policy outcomes. This financial power often translates and perpetuates inequalities, ensuring that the voices and interests of the economically powerful predominate over those of less affluent stakeholders and are given priority (Saaida, 2023). Such dynamics can lead to policy decisions that favour incremental changes or maintain the status quo, rather than adopting the bold measures necessary for significant environmental advancement (Ahlquist, 2017). It is unsurprising that conflicts of interest frequently emerge when economic development objectives clash with environmental protection efforts. This can be viewed in relation to the example of the aviation industry in Europe's objection to an increase in ETS in the region, as they viewed such as obstacles to their profitability and competitiveness without regard to the environmental benefits such would promote (Schvartzman, 2023). Even in governments, the pursuit of immediate economic benefits often overshadows long-term sustainability considerations, resulting in policy decisions that may aggravate environmental degradation (Čábelková et al., 2023). Identifying and reconciling any potential conflicts is a crucial step towards establishing a harmonious balance between economic ambitions and environmental imperatives. For sustainable futures to be realized successfully, it necessitates fostering collaboration and dialogue among the diverse array of stakeholders involved in the formulation of policies and frameworks.

Diverse strategies to encourage such collaboration need to be devised, to ensure that the aspirations of any of the groups are considered but are not given priority above that of the sustainability agenda. Potent strategies could include the establishment of multi-stakeholder forums, consensus-building processes, and the creation of integrated policy frameworks that accommodate the diverse needs and perspectives of all involved parties (Ansell & Gash, 2007). Through these collaborative endeavours, stakeholders can co-develop solutions that are not only innovative but also broadly acceptable, paving the way for more inclusive and effective sustainability initiatives.

Beyond local and national stakeholders and government, the role of international governance frameworks and transnational alliances is indispensable in addressing the global challenges of sustainability. This can be acknowledged by appreciating that environmental issues, by their nature, cross national boundaries and require a concerted international response. For instance, the impacts of climate change have been worldwide, with the least emitters, such as the Small Island Developing States (SIDS) and less developed economies shouldering the greatest burden (Barnett et al., 2015). Initiatives like the Paris Agreement on climate change demonstrate the potential of global cooperation to secure commitments from nations towards reducing greenhouse gas emissions and investing in sustainable development practices (Government of the United States, 2021). Furthermore, transnational alliances, which include collaborations between countries, NGOs, and the private sector, are instrumental in sharing best practices and pooling resources to support sustainability efforts on a global scale.

6 Shaping Political Dynamics

In summing up the above discussion, understanding the complex interplay between public opinion, media coverage, and cultural narratives is crucial in shaping the political landscape of sustainability issues. This is because each of these players, in one way or another leverages framing strategies to influence public perceptions and policy debates, whether for the benefit of all or for their own interest. Understanding the complexities involved is critical, especially in exposing the impact of misinformation and propaganda that emanates from diverse groups as strategies to achieve their objectives, and the potential of alternative media and citizen journalism in promoting sustainability.

In the discussion above, it has been widely covered how different actors use framing strategies to influence public perceptions and steer policy debates on sustainability. For instance, it is not surprising how the acts craftily highlight certain aspects of sustainability issues while downplaying others to shape how the public understands and engages with these challenges. As such, the effectiveness of framing strategies in altering public opinion necessitates a critical examination of the narratives promoted by various stakeholders, whether governmental bodies, NGOs, corporations, and media outlets. Often, the spread of misinformation, use of propaganda, vested interests, and corporate greenwashing are pursued by an actor or actors involved in the sustainability discourse to sway or invoke emotion in the public, away from pertinent sustainability agenda; hence, representing significant obstacles to public understanding of those issues. These practices have the potential to distort public discourse, leading to confusion and apathy among citizens. For instance, in the recent past, the 15-Minute City concept, which champions the redesigning of cities to make it possible for residents to accomplish most of their daily activities by walking or cycling for only 15 minutes or less has been a recipient of increased misinformation (Allam et al., 2023c, 2023e). It was falsely claimed that the model would restrict people's freedom of movement, concentrating them in 15-minute zones; hence, fuelling fear of government control (Stanford, 2023; Allam et al., 2023a; Dixit & Allam, 2023). Analysing the sources and impacts of misinformation and greenwashing is essential for developing strategies to counteract their effects and promote accurate information regarding sustainability challenges and solutions.

A relief to the pro-sustainability (or pro-sustainability movement) has been due to the emergence of alternative media platforms and citizen journalism, which have proven to be significantly powerful tools for challenging dominant narratives and bringing attention to marginalized voices in the sustainability discourse (Issawi, 2021). These media outlets are providing platforms for diverse perspectives and in-depth analysis, thus playing a pivotal role in educating the public, fostering critical thinking, and mobilizing grassroots support for sustainability initiatives. Of importance to this are strategies for building public support and mobilizing grassroots action to advance sustainability initiatives. This involves not only raising awareness but also engaging citizens in meaningful ways that encourage active participation in sustainability efforts. The alternative media platforms and citizen journalism are making it possible to expose the role of community-led initiatives, public demonstrations, and online campaigns in galvanizing public support and influencing policy decisions warrants thorough exploration.

In view of these two citizen alternatives to mainstream media, it is now evident that fostering critical thinking, enhancing media literacy, and promoting civic engagement is foundational for empowering citizens as active participants in the quest for sustainable futures. By equipping individuals with the skills to critically evaluate information, engage with complex sustainability issues, and participate effectively in democratic processes, society can cultivate a more informed and proactive citizenry capable of contributing to meaningful change.

References

Abdelfattah, L., Deponte, D., & Fossa, G. (2022). The 15-minute City: Interpreting the Model to Bring Out Urban Resiliencies. *Transportation Research Procedia, 60*, 330–337.

Act CRJTV, Equal Employment Opportunities. (1964). Civil Rights Act of 1964.

Ahlquist, J. S. (2017). Labor Unions, Political Representation, and Economic Inequality. *Annual Review of Political Science, 20*, 409–432.

Allam, Z. (2019). Identified Priorities for Smart Urban Regeneration: Focus Group Findings from the City of Port Louis, Mauritius. *Journal of Urban Regeneration Renewal, 12*, 376–389.

Allam, Z. (2020a) The First 50 Days of COVID-19: A Detailed Chronological Timeline and Extensive Review of Literature Documenting the Pandemic. In *Surveying the COVID-19 Pandemic its Implications* (p. 1–7). Elsevier.

Allam, Z. (2020b). The Second 50 Days: A Detailed Chronological Timeline and Extensive Review of Literature Documenting the COVID-19 Pandemic from day 50 to day 100. In *Surveying the COVID-19 Pandemic its Implications* (p. 9). Elsevier.

Allam, Z. (2020c). The Third 50 Days: A Detailed Chronological Timeline and Extensive Review of Literature Documenting the COVID-19 Pandemic From Day 100 to Day 150. In *Surveying the Covid-19 Pandemic its Implications* (p. 41). Elsevier.

Allam, Z., Chabaud, D., Gall, C., Pratlong, F., & Moreno, C., et al. (2023a). Chapter 5—Exploring the Relationship of Time Keeping and Urban Morphology within the Economic Renaissance and the Postmodern Era. In Z. Allam, D. Chabaud, & C. Gall (Eds.), *Resilient and Sustainable Cities* (pp. 77–91). Elsevier.

Allam, Z., Chabaud, D., Gall, C., Pratlong, F., & Moreno, C. (2023b). Chapter 6—Enter the 15-Minute City: Revisiting the Smart City Concept

Under a Proximity Based Plan In Z. Allam, D. Chabaud, C. Gall, et al. (Eds.), *Resilient and Sustainable Cities* (pp. 93–105). Elsevier.

Allam, Z., Chabaud, D., Gall, C., Pratlong, F., & Moreno, C. (2023c). Chapter 7—On Proximity-Based Dimensions and Urban Planning: Historical Precepts to the 15-minute City. In In Z. Allam, D. Chabaud, C. Gall, et al. (Eds.), *Resilient and Sustainable Cities* (pp. 107–119). Elsevier.

Allam, Z., Chabaud, D., Gall, C., Pratlong, F., & Moreno, C., et al. (2023d). Chapter 8—Financing the 15-Minute City Concept and Its Infrastructural Ecosystem in Developing Nations through Fiscal Mechanisms. In Z. Allam, D. Chabaud, & C. Gall (Eds.), *Resilient and Sustainable Cities* (pp. 121–133). Elsevier.

Allam, Z., Chabaud, D., Gall, C., Pratlong, F., & Moreno, C., et al. (2023e). Chapter 9—Redefining Investable Infrastructure in Developing Nations in a Postpandemic Era: The Case of the 15-Minute City. In Z. Allam, D. Chabaud, & C. Gall (Eds.), *Resilient and Sustainable Cities* (pp. 135–146). Elsevier.

Allam, Z., Cheshmehzangi, A., & Jones, D. S. (2023f). Redefining Climate Policy and Economic Resilience in the Consuming World. In *Climate and Social Justice: The Political Economy of Urban Resilience and Mercantilism.* (pp. 77–102.) Springer Nature Singapore

Allam, Z., Cheshmehzangi, A., & Jones, D. S. (2023g). Climate and Social Justice in the Political Landscape of Urban Resilience and Mercantilism: A Concluding Note. In *Climate and Social Justice: The Political Economy of Urban Resilience and Mercantilism* (pp. 103–110). Springer Nature Singapore.

Allam, Z., & Newman, P. J. S. C. (2018). Economically Incentivising Smart Urban Regeneration. Case Study of Port Louis, Mauritius. *Smart Cities, 1*, 53–74.

Allam, Z., Nieuwenhuijsen, M., Chabaud, D., & Moreno, C. (2022a). The 15-Minute City Offers a New Framework for Sustainability, Liveability, and Health. *The Lancet, 6*, e181–e183.

Allam, Z., & Raisah Takun, Y. (2022). *Chapter 4: Smart Cities as an Urban Regeneration Avenue: Redefining the Efficiency and Performance of Cities Rethinking Smart Cities* (pp. 66–87). Edward Elgar Publishing.

Allam, Z., Sharifi, A., Giurco, D., & Sharpe, S. A. (2021). On the Theoretical Conceptualisations, Knowledge Structures and Trends of Green New Deals. *Sustainability, 13*, 12529.

Allam, Z., Sharifi, A., Giurco, D., & Sharpe, S. A. (2022b). Green New Deals Could Be the Answer to COP26's Deep Decarbonisation Needs. *Sustainable Horizons, 1*, 100006.

Ansell, C., & Gash, A. (2007). Collaborative Governance in Theory and Practice. *Journal of Public Administration Research and Theory, 18*, 543–571.

Awan, U., & Sroufe, R. (2022). Sustainability in the Circular Economy: Insights and Dynamics of Designing Circular Business Models. *Applied Sciences, 12*, 1521.

Barnett, J., Evans, L. S., Gross, C., et al. (2015). From Barriers to Limits to Climate Change Adaptation: Path Dependency and the Speed of Change. *Ecology and Society*, 20.

Berry, J. M. (2015). *Lobbying for the People: The Political Behavior of Public Interest Groups*. Princeton University Press.

Beunen, R., & Patterson, J. J. (2019). Analysing Institutional Change in Environmental Governance: Exploring the Concept of 'Institutional Work.' *Journal of Environmental Planning and Management, 62*, 12–29.

Bosone, M., Casola, M., Daldanise, G., & Vito, D. (2023). Stimulating Circular Urban Regeneration through Cultural and Sustainable Communities: The Proposal for a Green Blue Youth Vision 2030. *Sustainability, 15*, 11294.

Brundtland, G. H. (1987). Our common future: Report of the World Commission on Environment and Development. Oxford University.

Buchter, L. (2023). Addressing Racism and Islamophobia Under the Rules of Colorblindness: When Social Movements Engage in Category Work to Reform the Meanings of Regulatory Categories. *Strategic Organization, 21*, 149–185.

Čábelková, I., Smutka, L., Mareš, D., Ortikov, A., & Kontsevaya, S. (2023). Environmental Protection or Economic Growth? The Effects of Preferences for Individual Freedoms. *Frontiers in Environmental Science, 11*, 1129236.

Caiani, M. (2023). Framing and Social Movements. *Discourse Studies, 25*, 195–209.

Calhoun, C., & Fong, B. Y. (2022). Introduction. In C. Calhoun & B. Y. Fong (Eds.), *The Green New Deal and the Future of Work* (pp. 1–20). Columbia University Press.

Cheshmehzangi, A. (2020). The Analysis of Global Warming Patterns from 1970s to 2010s. *Atmospheric and Climate Sciences, 10*(3), https://doi.org/10.4236/acs.2020.103022

Cheshmehzangi, A., & Chen, H. (2021). *China's Sustainability Transitions: Low Carbon and Climate-Resilient Plan for Carbon Neutral 2060*. Springer.

Cheshmehzangi, A., & Dawodu, A. (2019a) *Sustainable Urban Development in the Age of Climate Change—People: The Cure or Curse*. Palgrave Macmillan.

Cheshmehzangi, A., & Dawodu, A. (2019b). Stakeholder Methods to Climate Change and Sustainable Development. In *Sustainable Urban Development in the Age of Climate Change—People: The Cure or Curse* (pp. 23–68). Palgrave Macmillan.

Cheshmehzangi, A., & Dawodu, A. (2019c). Case Study Reviews: People, Perspective, and Planning, In *Sustainable Urban Development in the Age of Climate Change—People: The Cure or Curse* (pp. 69–131). Palgrave Macmillan.

Cheshmehzangi, A., Dawodu, A., & Sharifi, A. (2021). *Sustainable Urbanism in China*. Routledge.

Cheshmehzangi, A., & Tang, T. (2022). *China's City Cluster Development in the Race to Carbon Neutrality*. Springer.

Cheshmehzangi, A., You, N., Siri, J., & Birch, E. (2024). *Harnessing Urban Innovation to Unlock the Sustainable Development Goals*. Springer.

Christens, B., Gupta, J., & Speer, P. (2021). Community Organizing: Studying the Development and Exercise of Grassroots Power. *Journal of Community Psychology, 49*, 3001–3016.

Cress, D. M., & Snow, D. A. (2000). The Outcomes of Homeless Mobilization: The Influence of Organization, Disruption, Political Mediation, and Framing. *American Journal of Sociology, 105*, 1063–1104.

Daniels, B., Follett, A. P., & Davis, J. J. H. L. (2019). The Making of the Clean Air Act. *Hastings LJ, 71*, 901.

Davidson, C. (1992). The Voting Rights Act: A Brief History. *Voting Rights Act of, 1965*, 7.

della Porta, D., & Diani, M. (2006). *Social Movements: An introduction*. Blackwell Publishing.

Diani, M. (1992). *The Concept of Social Movement., 40*, 1–25.

Dixit, A., & Allam, Z. (2023). Urban Mobility and Time Geography Within the Smart City Narrative. In Z. Allam Z (Ed.), *Sustainable Urban Transitions: Research, Policy and Practice* (pp. 189–200). Springer Nature Singapore.

Dsouza, L., Patil, A., & Diwan, C. (2021). Causes, Effects, and Possible Remedies for Deforestation in the Aarey Region. *National Journal of Professional Social Work, 22*, 22–29.

Eren-Erdoğmuş, İ, & Ergun, S. (2017). The Impact of Social Media on Social Movements: The Case of Anti-Consumption. In S. Gordon (Ed.), *Online Communities as Agents of Change and Social Movements* (pp. 224–252). IGI Global.

Fosu-Mensah, B., & Yirenya-Tawiah, D. (2014). Assessing Public Perception of Beach Quality for Ecotourism Development: A Case Study in Kokrobite Beach in the Greater Accra Region, Ghana. *Journal of Environment and Earth Science, 4*, 58–66.

Fridays for Future. (2018). *What We Do*. Available at: https://fridaysforfuture.org/.

Government of United States. (2021) *Paris Climate Agreement*. Available at: https://www.whitehouse.gov/briefing-room/statements-releases/2021/01/20/paris-climate-agreement/.

Hadden, J. (2015). Networks in Contention: The Divestment Movement Against Apartheid. *American Journal of Sociology, 121*, 1015–1057.

Heaps, D. (2019). *Sami land Rights and Policy-Driven Recognition Threats*. Available at: https://jsis.washington.edu/news/sami-land-rights-and-policy-driven-recognition-threats/.

Hiraide, L. A., & Evans, E. Intersectionality and Social Movements: A Comparison of Environmentalist and Disability Rights Movements. *Social Movement Studies*, 1–18.

Honeybun-Arnolda, E., Turner, R. A., Mukhopadhyay, R., Collins, C., & Wills, J. (2024). Localising and Democratising Goal-Based Governance for Sustainability. *Environmental Science & Policy, 151*, 103638.

IPCC. (2023). Climate Change 2023: Synthesis Report. Contribution of Working Groups I, II and III to the Sixth Assessment Report of the Intergovernmental Panel on Climate Change (Core Writing Team, H. Lee & J. Romero, Eds., pp. 35–115). IPCC.

Issawi, Fe. (2021). Alternative Public Spaces in Hybrid Media Environments: Dissent in High Uncertainty. *Journalism & Mass Communication Quarterly, 98*, 923–942.

Jasper, J. M. (2018). *The Emotions of Protest*. University of Chicago.

Kaiser, M. S. (2020). Are Bottom-Up Approaches in Development More Effective than Top-Down Approaches? *Journal of Asian Social Science Research, 2*, 91–109.

Kim, Y.-j, & Yang, H. J. (2023). Rethinking Cheonggye Stream Restoration Project: Is Urban Greening Strategy Socially Inclusive? *Land Use Policy, 131*, 106742.

Larmer, M. (2010). Social Movement Struggles in Africa. *Review of African Political Economy, 37*, 251–262.

Mccarthy, G. (2021). *Ayana Elizabeth Johnson*. Available at: https://time.com/collection/time100-next-2021/5937679/ayana-elizabeth-johnson/

Moreno, C., Allam, Z., Chabaud, D., Gall, C., & Pratlong, F. (2021). Introducing the "15-Minute City": Sustainability, Resilience and Place Identity in Future Post-pandemic Cities. *Smart Cities, 4*, 93–111.

Mosse, D., & Nagappan, S. B. (2021). NGOs as Social Movements: Policy Narratives, Networks and the Performance of Dalit Rights in South India. *Development and Change, 52*, 134–167.

National Museum of the American Indian. (2018). *Treaties Still Matter: The Dakota Access Pipeline*. Available at: https://americanindian.si.edu/nk360/plains-treaties/dapl

Newell, J., Foster, A., Borgman, M., & Meerow, S. (2022). Ecosystem Services of Urban Agriculture and Prospects for Scaling up Production: A Study of Detroit. *Cities, 125*, 103664.

Rahman, M. F., Falzon, D., Robinson, S.-a, et al. (2023). Locally Led Adaptation: Promise, Pitfalls, and Possibilities. *Ambio, 52*, 1543–1557.

Regional Health-Americas, T. L. (2022). Indigenous Peoples' Territorial Sovereign in the Amazon must be Respected. *Lancet Regional Health-Americas, 11*, 100331.

Rifkin, J. (2019). *The Green New Deal: Why the Fossil Fuel Civilization Will Collpase by 2028, and the Bold Economic Plan to Save Life on Earth*. Martin's Press.

Roberts, A. (2012). Why the Occupy Movement Failed. *Public Administration Review, 72*, 754–762.

Roth, S. (2021). Intersectionality and Coalitions in Social Movement Research—A Survey and Outlook. *Sociology Compass, 15*, e12885.

Saaida, M. (2023, December 1). The Influence of Money in Politics, 1–12. https://www.researchgate.net/publication/375237736_The_Influence_of_Money_in_Politics

Schlosberg, D., & Collins, L. B. (2014). From Environmental to Climate Justice: Climate Change and the Discourse of Environmental Justice. *Wiley Interdisciplinary Reviews: Climate Change, 5*, 359–374.

Schmid, N., Sewerin, S., & Schmidt, T. S. (2020). Explaining Advocacy Coalition Change with Policy Feedback. *Policy Studies Journal, 48*, 1109–1134.

Schvartzman, R. (2023). *EU ETS Reform Destabilizes International Consensus for Aviation Carbon Reductions*. Available at: https://www.iata.org/en/about/worldwide/europe/blog/eu-ets-reform-destabilizes-international-consensus-for-aviation-carbon-reductions/

Skaidrė, Ž. (2019). Grassroots Activism and Sustainable Development. In W. Leal Filho (Ed.), *Encyclopedia of Sustainability in Higher Education* (pp. 1–10). Springer International Publishing.

Stanford, C. (2023). *The 15-Minute City: Where Urban Planning Meets Conspiracy Theories*. Available at: https://www.nytimes.com/2023/03/01/world/europe/15-minute-city-conspiracy.html

Sun, L., Chen, J., Li, Q., & Huang, D. (2020). Dramatic Uneven Urbanization of Large Cities Throughout the World in Recent Decades. *Nature Communications, 11*, 5366.

Sunrise Movement. (2023). *Our Story: Sunrise Movement*. Available at: https://www.sunrisemovement.org/theory-of-change/our-story/

Tischner, A., & Stasiuk, K. (2023) Spare Parts, Repairs, Trade Marks and Consumer Understanding. *IIC—International Review of Intellectual Property and Competition Law, 54*, 26–60.

Tyler-Davies, M. (2019) *A New Fossil Free Milestone: $11 Trillion has been Committed to Divest from Fossil Fuels*. Available at: https://350.org/11-trillion-divested/#:~:text=Over%201110%20institutions%20have%20now,universities%2C%20faith%20groups%20and%20foundations

UNFCCC. (2020). *Joint Statement on the US Withdrawal from the Paris Agreement*. Available at: https://unfccc.int/news/joint-statement-on-the-us-withdrawal-from-the-paris-agreement

United Nations. (2020). *United Nations Sustainable Development Goals*. Available at: https://sdgs.un.org/goals

Van Dyke, N., & McCammon, H. J. (2010). *Strategic Alliances: Coalition Building and Social Movements*. University of Minnesota Press.

Vasoo, S. J. C. D. J. (1991). Grass-root Mobilisation and Citizen Participation: Issues and Challenges. *Community Development Journal, 26*, 1–7.

Whyte, K. (2018). Settler Colonialism, Ecology, and Environmental Injustice. *Environment and Society, 9*, 125–144.

Woo, E.-J., & Kang, E. (2020). Environmental Issues As an Indispensable Aspect of Sustainable Leadership. *Sustainability, 12*, 7014.

Yoshioka, T. (2010). Lobbying. In H. K. Anheier & S. Toepler (Eds.), *International Encyclopedia of Civil Society* (pp. 950–955). Springer.

Zhu, Y., Zhang, H., Siddik, A. B., Zheng, Y., & Sobhani, F. A. (2023). Understanding Corporate Green Competitive Advantage through Green Technology Adoption and Green Dynamic Capabilities: Does Green Product Innovation Matter? *Systems, 11*, 461.

CHAPTER 5

The Green New Deal as a Pathway for Climate and Social Justice

Abstract The Green New Deal (GND) represents a multifaceted strategy aimed at addressing climate change and promoting social justice. Over the years, scholarly interest in climate action has surged, with substantial literature on the GND emerging even before its formal adoption in the U.S. in 2019. This chapter utilizes data from a bibliometric analysis to navigate the extensive body of work on the GND, providing a comprehensive overview of research trends and thematic focus areas from 1995 to 2021. The analysis highlights three significant periods, each marked by unique developments in the discourse on sustainability, economic growth, and climate policy. From 1995 to 2014, key terminologies such as climate change, sustainability, and green growth were central to academic discussions. The period from 2015 to 2019 saw a significant increase in literature, influenced by global initiatives like the Paris Agreement and the Sustainable Development Goals. The COVID-19 pandemic further intensified research on the GND, with new terms such as green recovery and energy transition gaining prominence from 2020 to 2021. The chapter underscores the GND's potential to drive climate justice through three key dimensions: addressing historical injustices, promoting equitable solutions, and linking global climate action with local struggles. It emphasizes the importance of inclusive decision-making, reskilling and upskilling workers, and fostering international cooperation to achieve a sustainable and just transition. The bibliometric analysis provides valuable

Z. Allam and A. Cheshmehzangi, *Sustainable Futures and Green New Deals*, https://doi.org/10.1007/978-3-031-63642-4_5

insights for researchers and policymakers to craft effective and contextually appropriate GND models, ultimately contributing to the global effort to combat climate change and promote social equity.

Keywords Green New Deal (GND) · Climate justice · Sustainability · Renewable energy · Economic growth · Social equity · Bibliometric analysis · Policy frameworks

1 Introduction

Over the years, the focus on climate action has been gaining popularity among scholars, institutions, and civil societies concerned with how the climate agenda has been evolving. As a result, they have voiced their concerns through a wide range of publications and literature, and as a result, the world is now amassed with diverse knowledge on the topic. Regarding GNDs, substantial studies and research have been done, even before the first official version was adopted in the United States in 2019. The trigger to these publications has been the negative impacts associated with climate change, and how government and relevant bodies have been unable to propose long-lasting solutions. Research and studies were also prompted by the increasing opportunities for countries to continue adopting unsustainable approaches in their endeavours such as economic development.

The rapid increase in literature on GNDs, and the need to understand the trends in the recent past, especially after the COVID-19 pandemic, poses a challenge, especially to those eager to explore the concept of GND (Allam et al., 2021b). However, the challenge can be overcome by utilizing bibliometric analysis techniques that allow gaining an overview of specific fields. Such macroscopic overviews can be used to highlight major thematic focus areas and research trends. It should be understood that various bibliometric analysis software tools such as SciMAT, CiteSpace, and VOSviewer have been developed in the past two decades.

Analysing different publications and literature is important, particularly post-COVID-19, and with the ongoing global challenges such as Russian-Ukraine and Israel-Palestine war, which have the potential to undermine the climate actions already in place. Further, as was observed regarding the distribution of the COVID-19 vaccine, the levels of inequality and injustices continue to increase, and if solutions are not well spelled out even in the GNDs, chances of failure are imminent. Therefore, by understanding the past, it could be possible to make right the future. It is on this premise that this chapter is anchored.

2 Methodology Description

Borrowing from a previously published studies (Allam et al., 2021b), the authors sourced data for a bibliometric analysis using VOSviewer from academic databases such as Scopus and the Web of Science (WoS). Using this search string in the WoS on August 5, 2021, returned 1403 articles, where only 1174 articles were selected for final analysis in the VOSviewer.

As shown in Fig. 1, there has been a surge in the number of publications around 2015. This could be linked to the introduction and adoption of international policy frameworks such as the Paris Climate Agreement and the 2030 Agenda for Sustainable Development in 2015. Accordingly, 2015 was designated as the first milestone. The figure also shows another surge in 2020 which could be linked to the emergence of the COVID-19 pandemic. Based on these milestones, the study period was divided into three sub-periods, namely, 1995–2014, 2015–2019, and 2020–2021. Outputs of all VOSviewer analyses are presented as a network of nodes and links. The node size is an indication of the relative importance. For instance, in the case of term co-occurrence analysis, node size is proportional to the frequency of term occurrence. Also, link width is proportional to the strength of the connection between two terms. Terms that are closely linked to each other form clusters that indicate thematic focus areas.

3 Term Co-Occurrence

3.1 All Periods

Results of the term co-occurrence analysis for the whole study period are shown in Fig. 2. The terms were categorized and clustered into three groups. The Red cluster categorizes terms associated with climate change and sustainability, while the blue cluster encompasses terms associated with green growth. The Green cluster was dedicated to terms associated with the application of technologies and policies to foster economic growth. The Red cluster had a total of 25 terms, with key

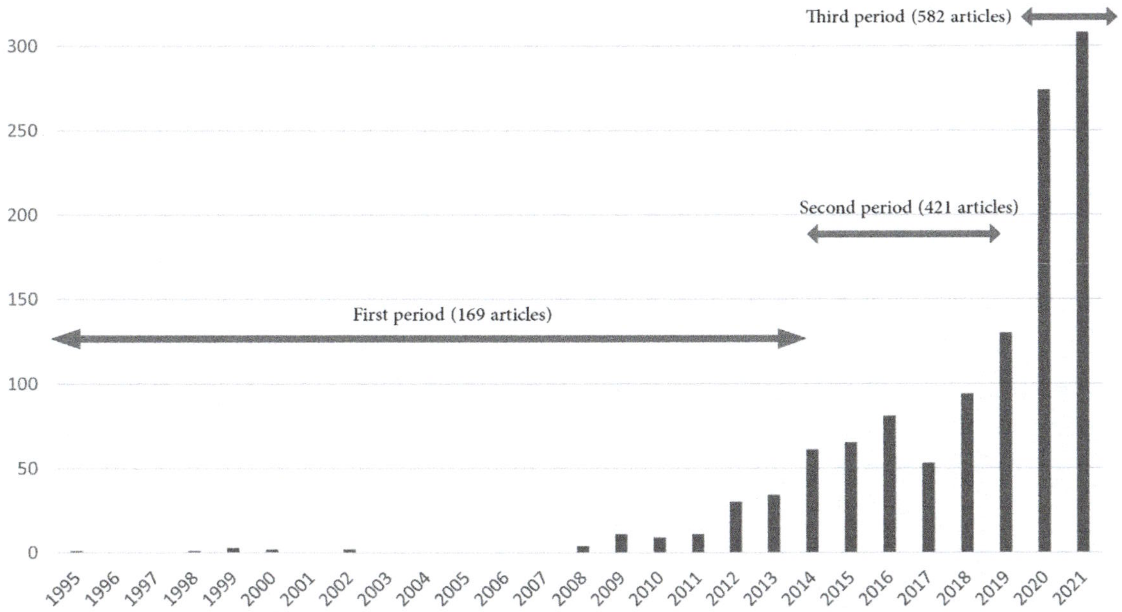

Fig. 1 The number of articles published per year. Note that although the literature search was conducted in August 2021, the number of articles in this year is already larger than its the previous year. This is a clear indication of increasing interest in this topic and more articles are expected to be published in the rest of 2021 and coming years, especially in view of the impacts of COVID-19 on many economies and the need for prompt recoveries aligning with both economic and sustainability pursuits (Generated by the authors using data from: Allam, 2020a, 2020d; Allam & Jones, 2020)

terms (according to the size of their node) including sustainability, which was deemed to be more closely associated with terms like green growth and green economy from the blue cluster. The term climate change was almost centrally positioned as shown in Fig. 8 and is seen to be associated with terminologies from both the green and blue clusters. The Green cluster had a total of 15 terms, with policy, economic growth, and efficiency being the main ones and seemed to be closely linked with other terms in this cluster and from the blue and red clusters as well. Other terms that were popular in this category include efficiency, innovation, impact, technology, renewable energy, China, and CO_2. The blue cluster had only six terminologies, with the key one being green growth, which is seen to be closely associated with major terminologies like sustainability, climate change, energy, Green New Deal (GND), and others from the red cluster. Likewise, it has close links with terms like *Economic Growth, Impacts, Model Performance, Innovation,* and *Efficiencies* from the Green cluster.

When studying the terms more closely in three distinct periods, we are offered with interesting insights as to their evolution across time. This is expanded in the 3 sub-sections below.

3.2 First Period (1995–2014)

During the period 1995–2014, a number of terminologies related to the topic of GND, as shown in Fig. 3, were already popular with most academicians. A total of 34 key terminologies were identified during this period, classified into three clusters; Red—representing sustainability and climate change, blue cluster encompassing terms like *Green Growth, Developing Countries* and others commonly associated with economic growth and development. The Green cluster with 14 terms is observed to focus on technology, environment, and innovations. The key terminologies during this period included *Green Growth, Sustainability, Climate Change, Policy* and *Energy*. However, other geographies that had begun attracting substantial attention during this period include China and South Korea, the latter aligning with the findings of the geographical popularity as represented in Fig. 6, where South Korea was noted as being the region attracting the most interest for GNDs. It is further noted that those two regions showed a particularly high interest in the terms *Green Growth, Growth, Technology, and Environmental Sustainability.* In the sustainability and climate change discourse, terms that

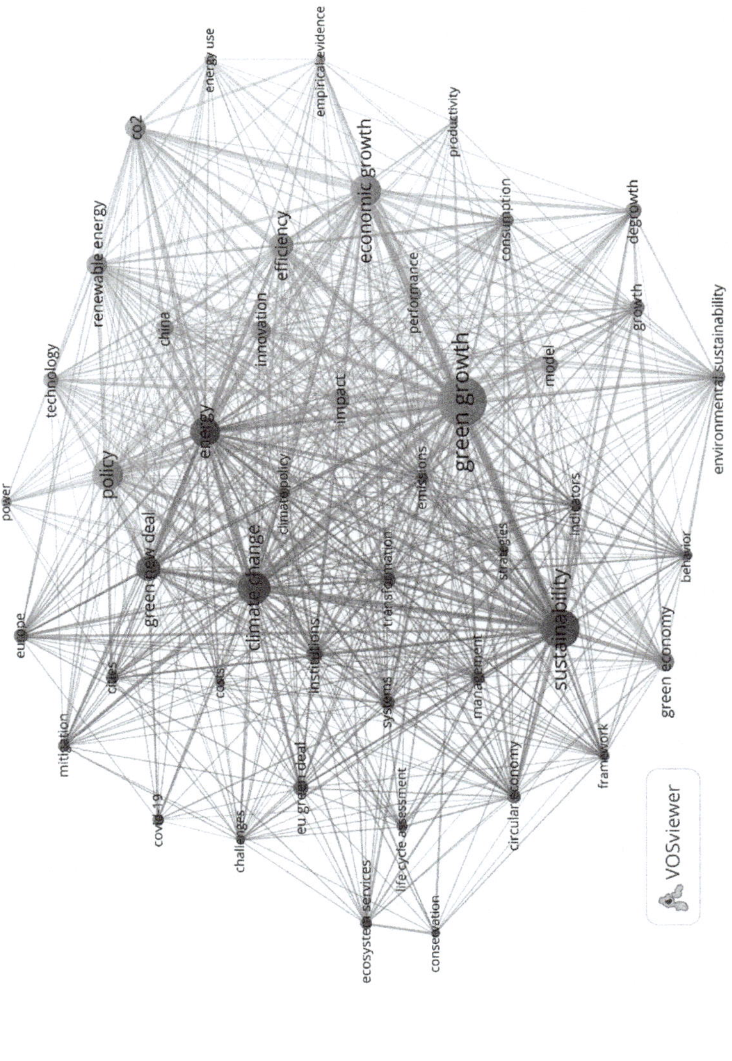

Fig. 2 Analysis of terms co-occurrence from 1995 to 2021. Illustration generated by the Authors, using VOSviewer

gained popularity include *Renewable Energy, Institutions, Management,* and others. Regarding technology and innovations, terms that were seen to attract attention (Cheshmehzangi et al., 2024), but were not yet linked to sustainability and economic growth include housing, efficiency, innovations, and others.

From the analysis, it is not surprising that already, terms like *Climate Change, Green Growth, Sustainability, Policies* and *Energy* were already popular and co-occurring in discourses touching the three thematic areas related to GND. Within this period, as expressed by FAS, countries like China and South Korea were experiencing unprecedented growth, courtesy of technology and innovations that were popular in those countries. The growth in China began in the 1970s, and by the 1990s, had become a major trade destination, attracting over 445,244 Foreign Direct Investments (FDIs) businesses in the 1990s (Congressional Research Service, 2019). The increase in trade was inspired by the deployment of advanced technologies in different sectors. In Korea, which is also seen to have drawn the attention of authors, from 1995 onwards, its growth is observed to have recovered from an unprecedented decline at the beginning of the 1990s (Congressional Research Service, 2019). It is reported that the country had embarked on technological deployment in different sectors, and by 1996, the country had started to grow by 7.8% (O'Neill, 2023).

Growth was not only being experienced in this region alone, as noticed in other countries like the United States. Such economic activities across the globe were prompted by the recovery strategies that countries had embarked on as they tried to recover from the 1990s recession (Brian & Patrick, 2010). As a result, coupled with ongoing environmental pressures, usage of non-renewable energy was experienced in countries like China and the United States which had only 7% renewable energy with the rest coming from fossil fuel 24%, coal 23% and natural gas (23%) (Joskow, 2001). As a result, concerns about climate change intensified, prompting the formulation of the Kyoto Protocol that came into force in 1997 (United Nations Framework Convention on Climate Change, 1998). The economic growth that was being experienced prompted the emergence of housing programs, especially low-cost housing in many developing regions, especially in Asia and North America; as disposal income following the increase in wages started to rise (The World Bank, 2022). Through innovations, affordable housing became the trend, and countries like the United States, Canada, Singapore, and others were

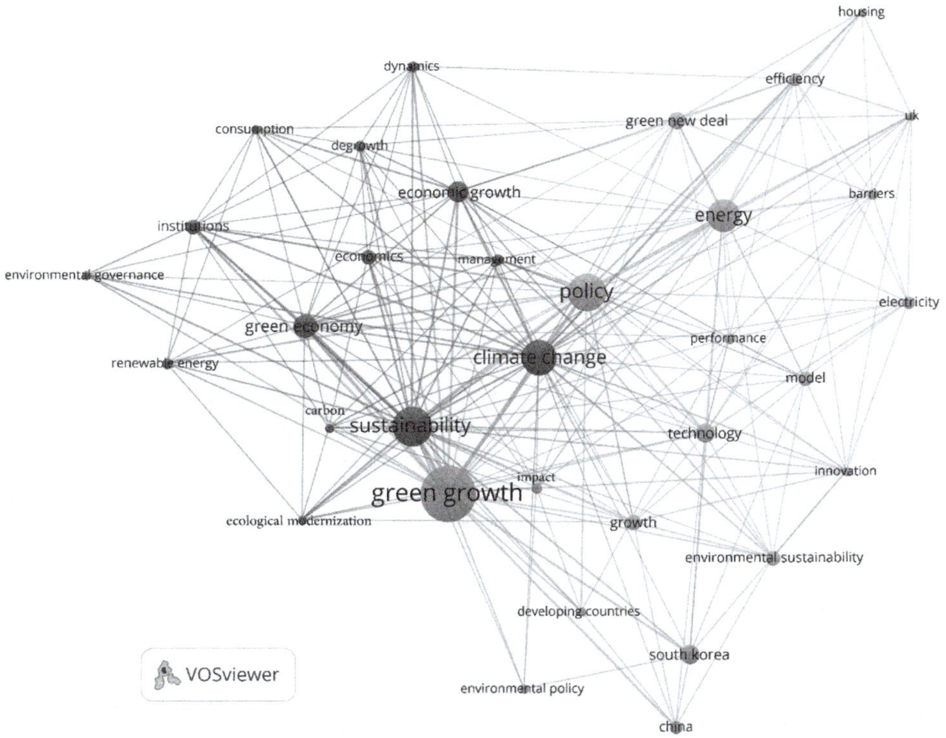

Fig. 3 Analysis of terms co-occurrence from 1995 to 2014. Illustration generated by the Authors, using VOSviewer

in a rush to ensure their citizens, whose numbers were also increasing, especially in urban areas were housed (Evans, 2014). However, while pursuing housing projects, it is noted that concerns arise regarding energy efficiency and the need for the adoption of alternative clean energy. Additionally, it is noted that the 'green' concept was still not popular during this first period.

3.3 Second Period (2015–2019)

During this period, as depicted in Fig. 4, the number of terminologies increased significantly to cover emerging issues. These terms are categorized into three groups, represented by red, blue, and green clusters. The Red cluster captures all the terminologies directly associated with sustainability and climate change while the blue cluster is dedicated to all terminologies with a greater inclination to *Energy Production, Consumption,* and *Impacts*. The Green cluster encompasses terms focusing on *Green Growth*, as influenced by factors like *Technology, Policies, and Innovations*. During this period, it is evident that all the terms recorded in the first period (1995–2014) are still present here and have been reinforced by the emergence of new terms. In the red category, it is evident that new terms such as *Impacts, Cities, Power, Productivities, Transmission,* and others have become common, and are occurring in documents focusing on all three thematic areas. Similarly, in the green cluster which was presented as the blue cluster in the first period, in addition to the six terms that were recorded then, numerous more terms emerged during this second period. Notable new terms here include *Green Economy, Europe, Renewable Energy, Emissions, Scenarios Behaviours, SDGs,* and others. In the Blue cluster, the number of terms that emerged was not many compared to other clusters. They include *Innovation, Buildings, Retrofit,* and UK. The terms *Policy, Efficiency, Innovation, and Green New Deal* however are seen to have become slightly more popular during this period, and also highly linked with terms like *Energy, Renewable Energy, Technology, Impacts, Green Growth,* and others.

From the analysis, it is observed that this second period experienced major breakthroughs in the discourse on *Sustainability, Equitability, and Economic Growth*. It is during this period that the Paris Agreement (2015), Sustainable Development Goals (SDGs) 2015, the Agenda 2030 for Sustainable Development (2015), the New Urban Agenda (2016), the Green New Deal (American model) 2019 and others were

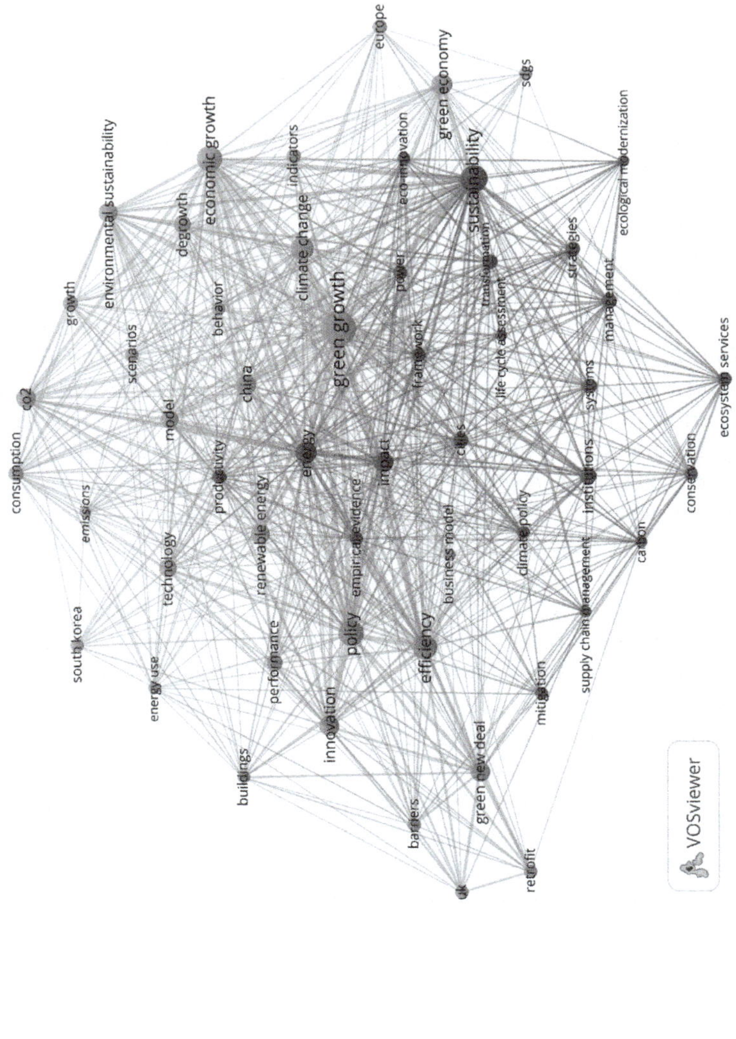

Fig. 4 Analysis of terms co-occurrence from 2015 to 2019. Illustration generated by the Authors, using VOSviewer

formulated, and where some are already in force. In fact, there have been numerous meetings on different agendas, especially on environment and sustainability during this period, with a large number of publications published, including notable documents such as the SDGs and the Paris Agreement. During this period, it became apparent that the world was experiencing increasing pressures from the (still) unprecedented challenges of climate change, impacting mostly vulnerable economies like Least Developed Countries (LDCs) and Small Island Developing Economies (SIDS) (UNEP, 2014). It is during this period also that sustainability agendas increased more with areas like urban planning being at the forefront, with urban models like Sustainable Smart City models being adopted in numerous geographies, including Songdo, South Korea (Allam, 2020f; Allam & Newman, 2018, 2023; Kolotouchkina & Seisdedos, 2017). Other cities include Singapore (Rohaidi, 2018), Barcelona (Gascó-Hernandez, 2018), and other parts of the world that have transformed, courtesy of the proliferation and ubiquitousness of smart technologies (Allam et al., 2022a; Chamoso et al., 2018; Samad et al., 2019; To et al., 2018). Such activities saw an increase in the academic arena with numerous co-occurring terminologies increasing and covering different aspects of *Sustainability, Growth, Economy, Climate Change, Technologies,* and more. With the global population increasing at an average of 140 million per year since 2015 as expressed by Roser et al. (2019), substantial changes in different global sectors were observed, especially as demand for services like energy, transport, manufactured products, and others increased. These then prompted the need for policy interventions to ensure that as consumption continued, identified themes like sustainability, *Conservation,* and *Supply Chain Management* were noted, as captured in the Paris Agreement and the New Urban Agenda (NUA) documents (United Nations, 2017).

3.4 Third Period (2020–2021)

Analysis of the period 2020–2021, as shown in Fig. 5, further highlighted the consistency of increased terminologies shaping the discourse on GND, within the three thematic areas of Sustainability and Climate change categorized under the red cluster. The second thematic area is energy and renewable energies highlighted in the blue cluster, while the third thematic area categorized under the green cluster encompassing terms focusing on *Green Growth, Economic Development, and Technologies*. The results here show that like in the previous periods (1995–2014 and 2015–2020), new terms arising from global trends emerged. These include *Green Recovery, COVID-19, Agriculture, Biodiversity, Bioeconomy, Circular Economy, EU-Green Deal, Storage, Photovoltaics, Energy Transition, GHGs, LCA, Determinants, Countries, Empirical Evidence, Panel Data, Environmental Kuznets Curve and Decoupling* among others. Some other terms that were common in the previous period but were no longer popular include *South Korea, Behavior, and Productivity*.

From the analysis, this period portrayed mixed results for the global environment, and the discourse on sustainability, as a result of the outbreak of COVID-19 in early 2020 (Allam, 2020a, 2020d). On the one hand, measures instituted in different countries and regions saw a reduction of activities in major sectors like tourism, transport, manufacturing, education, hotels and hospitality industry, and retail among others, resulting in a drastic reduction in emissions; by approximately 6.4% equivalent to 2.3 billion tonnes of CO_2 (Tollefson, 2021). However, the pandemic also resulted in an unprecedented plummeting of global economic growth; estimated at around 3.5% by IMF (Gopinath, 2021), more so in developed economies.

Thereafter, in the third quarter of 2020, countries started to ease COVID-19 restrictions and gradually started to strategize on how to get back their economies on track (Allam, 2020c, 2020e); thus, sparking fears that some would overlook their climate commitments and revert to the use of non-renewable energies (Allam & Jones, 2021; Allam et al., 2021a; Chabaud et al., 2023). This may explain why terms like *Energy Transition, Decarbonization, Photovoltaic Climate Policy, GHGs, Environmental Sustainability, Panel Data,* and others have emerged co-concurrently with the existing ones. On this, there have been calls for economies to consider embracing renewable energies in their COVID-19 recovery plans, with solar photovoltaic being championed, especially after notable advancements in the technologies used in making solar panels and other components like storage batteries helped reduce the prices thereof (Marcacci, 2020).

During this period also, the EU Green Deal was proposed (European Commission, 2019), with proposals of formally adopting the Circular Economy concept as a means to changing the economic profile of the EU region. According to Dartford (2021), this deal impacts the relationship between the EU and its trading partners, more so with respect to the

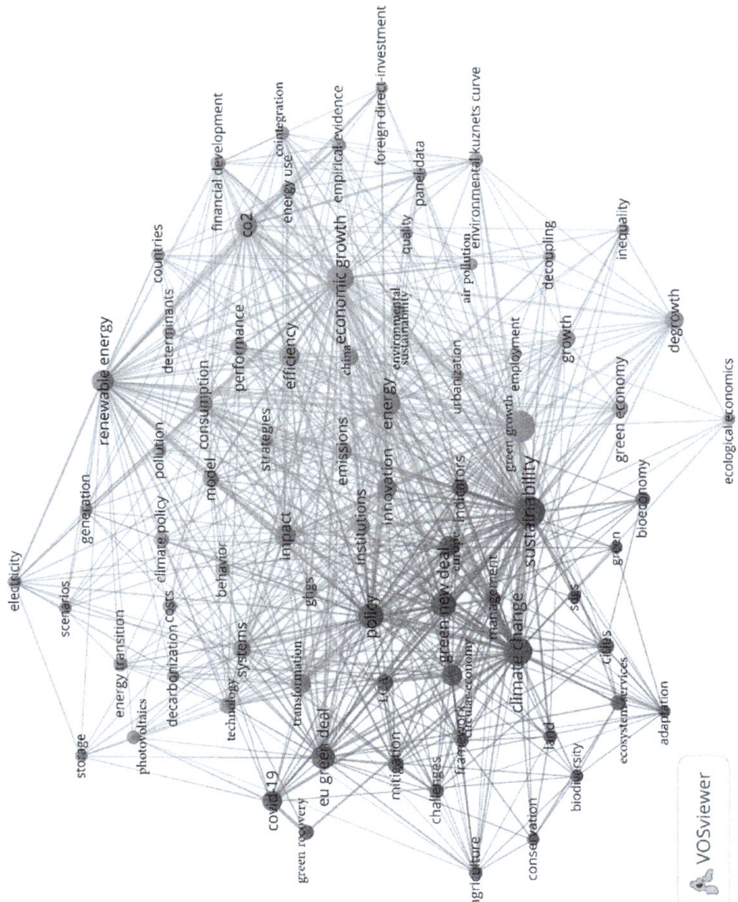

Fig. 5 Analysis of terms co-occurrence from 2020 to 2021. Illustration generated by the Authors, using VOSviewer

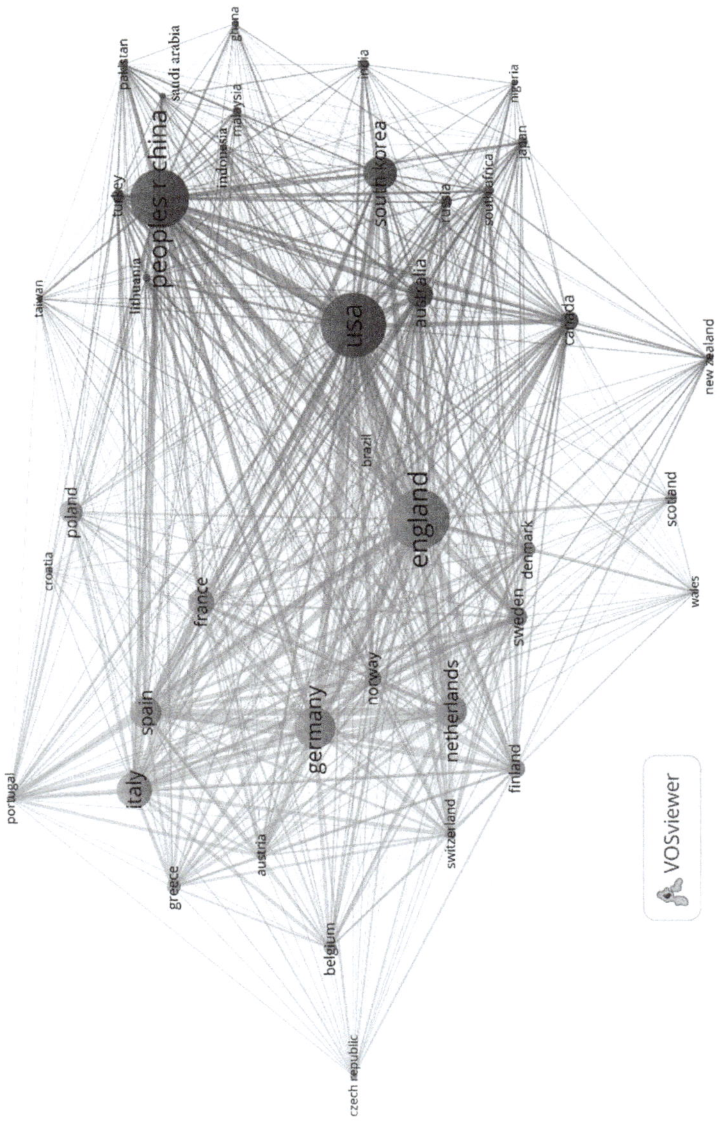

Fig. 6 Countries with the most contributions to GND. Illustration generated by the Authors, using VOSviewer

sustainability aspects of raw materials and products exported from and to Europe. The bottom line during this period was for economies to emphasize Green Growth, positing as a strong approach for positive impacts even for LCDs and SIDS, observed to be struggling financially because of the debt crisis exacerbated by the impacts of COVID-19 (Allam, 2020b; United Nations, 2021).

The focus is on implementing policy frameworks that would encourage economies to continue investing in sustainable pathways, like as renewable energy and increased use of technologies in areas like urban planning. This, as noted by the World Bank (2021) will promote a gradual economic recovery, while at the same time ensuring that the global 2030 targets for reduction in emissions are not overlooked. This would be important, even with the ranging political upheavals between countries such as Russia and Ukraine and Israel and Palestine. While it is important to consider and concentrate on ending these wars, the focus on a green, just, and equitable world should not be lost.

4 The GND'S Potential for Transformation

The discourse in this book is centred on the potential of GNDs, and how much has been embraced in different parts of the world as shown in Fig. 6 where the United States, the People's Republic of China, England, and Germany have been highlighted as the most popular contributors to the GND. This is evident by the size of the nodes, which in this case the four countries have the largest (as per the number of research documents emanating from those regions). However, the popularity of GND as noted in Chapter 1 has over the years also gained traction in countries like the Netherlands, France, Italy, South Korea, Australia, Spain, and Poland as depicted in Fig. 2.

It is worth noting that the results depicted in Fig. 6 are not surprising as the current global pressures towards the adoption of a Green Deal have been seen to concentrate more in the U.S. and European zones. For China, as noted by Bradshaw et al. (2021), though there is no formal agenda that can be equated to the Green New Deal, there has been an increase in attention towards activities pointing towards a reduction of carbon footprint. The increased attention and research focusing on China may be due to the fact that it is the leading emitter of carbon emissions, accounting for almost 24% of the global emissions (Bradshaw et al., 2021). Further, being the leading country in investment in renewable energy may have contributed to increased attention on its focus on matters relating to the adoption of deeper measures, aligning with the Green New Deal.

The attention in the United States also maybe multi-pronged. First, since the 1970s as reported by Conte (2019), there have been numerous calls, proposals, and agitations for the country to adopt climate actions to avert its impacts on the environment, economy, and communities. Secondly, as was expressed by Bradshaw et al. (2021), it is ranked second globally in terms of global emissions, and its commitment to global accords like the Paris Agreement, especially in view of the previous government policies on the same, may have triggered an increased interest. Thirdly, it is the first country to have a realistic Green New Deal formulated and discussed at the Congress level (Congress, 2019), thus drawing global attention. In the case of England, its contribution to the discourse on GND (or Green Deal as it is known in Europe) ranks it top among European countries. From the literature, it is noted that though the Green Deal will impact the entire European region and beyond, England, would be a major benefit due to its focus on areas like Green spaces and improvement of air quality that was noted to be fairly poor compared to its counterparts in the region (London Assembly, 2021).

The debate on GND is not only focused on countries but also on how reputable organizations such as learning institutions have been at the forefront in agitating for climate action, just society, and fairer economic approaches (Allam et al., 2021b, 2022b). Figure 7 showcases the contribution of different organizations regarding GND, more so through the number of publications each has generated. In brief, the Red cluster comprises 12 organizations with larger nodes that are closely linked noting the thickness of their link lines. The Blue cluster is located far right, with only three organizations (*Vrije University, University Autonoma Barcelona,* and *ICREA*) therein with each of them having large nodes and close links between themselves. However, this cluster is far linked from the rest of the clusters. The Green cluster comprised of 9 organizations that are closely linked to each other but with smaller nodes, indicating that the number of publications from those organizations was fairly few compared to those obtained in other clusters.

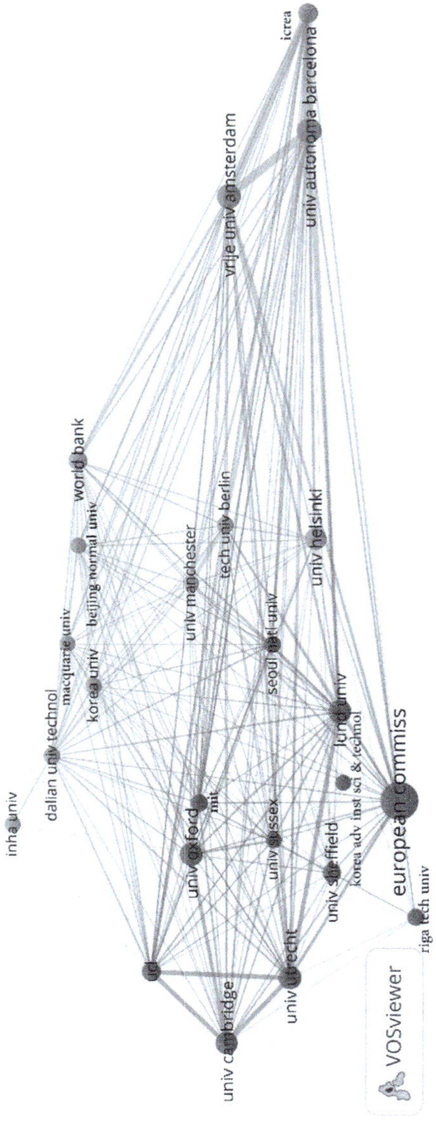

Fig. 7 Organizations contributing the most to GND. Illustration generated by the Authors, using VOSviewer

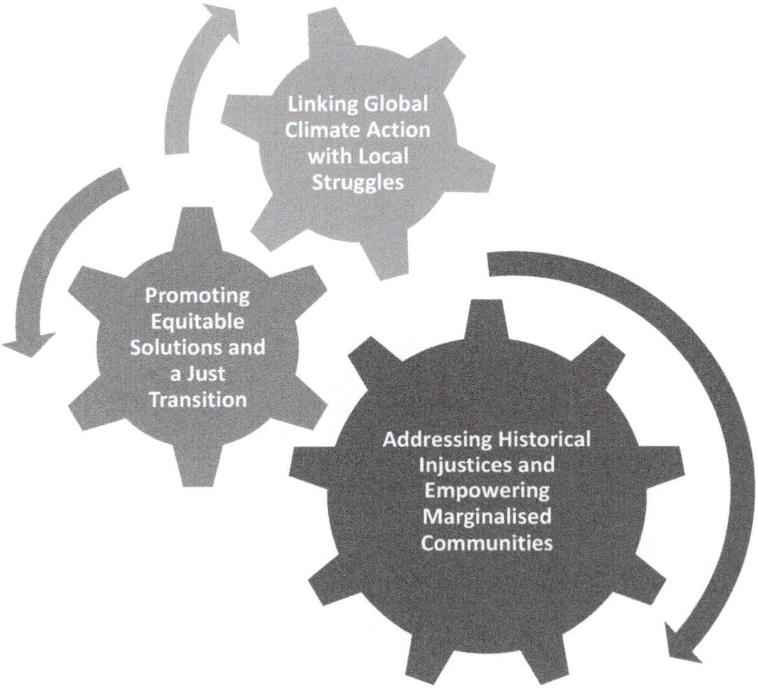

Fig. 8 The diverse role of the GND in achieving climate justice through three key dimensions

From the literature review, the *European Commission (EC)* has been identified as one of the organizations that eventually managed to bring to fruition the Green Deal agenda in Europe, and from the diagram above, this could be deduced to be due to its close link with numerous learning institutions. This explains why most of the documents that could have shaped the diverse GNDs associated with the region may have emanated from those institutions. This confirms that the European Green Deal emanated from different environmental knowledge shared, and exerted from different quarters in the region, especially from learning institutions. This is true following realization that, indeed, there are notable concerns with issues like increasing emission levels from the region (Bradshaw et al., 2021; González-Sánchez & Martín-Ortega, 2020), the unsustainability of the economy due to some sectors still relying on non-renewable energy sources (European Union External Action, 2023) and the increasingly observable cases of social inequalities in the region as document in a report by Eurofound (2023).

One clear observation from the results above is that most of the organizations are learning institutions drawn from the Euro-zone save for the few drawn from Korea, the World Bank, and the European Commission. From the analysis, the United States and Europe can be noted as the only two to have clear-cut GNDs. In America, the political class drawn from both the Democrats wing and other smaller parties have been credited for pushing for the realization of GND and this could be due to the fact that other institutions, such as NGOs and learning institutions are not as coordinated in research and policy work, as is the case with Europe. However, the results from Europe highlight the influence that learning institutions can have in shaping and influencing meaningful changes in society. Whereas it may be construed that the topic of GND is not widely understood in the public domain, hence, justifying why most organizations pushing for actions to be taken are learning institutions, evidence from the literature shows otherwise. On this, youth movements (Frank et al., 2018; Fridays For Future, 2021), UN bodies like the UNFCCC (United Nations Framework Convention on Climate Change, 1998, 2015; UNFCCC) and the IPCC (Intergovernmental Panel on Climate Change, 2014; IPCC, 2021), and other bodies, have been clear on their position on climate change. However, it is also important to appreciate that adopted GND models do not only concern the environment, but encompass other cross-cutting issues, that learning organizations, the World Bank, the EC, and other organizations keenly support (Walker, 2022). This showcases a global consensus on expanding the GND to touch on socio-economic dimensions, hence using the rationale for approaching the environment as an underlying foundation to regenerate societies.

The increasing attention accorded to diverse GNDs is justifiable and timely, as has been discussed in this document. Its focus on other aspects such as social justice and economic empowerment of vulnerable communities and communities cannot be overlooked. For instance, the proposals of formally adopting the Circular Economy concept as a means to changing the economic profile of the EU region is welcome as such creates unprecedented opportunities for job creation and economic empowerment as well as a shift to sustainable practices (Cheshmehzangi & Tang, 2022; Cheshmehzangi et al., 2021). However, it should be noted GNDs proposed have the potential to elicit friction between different countries and political players. For instance, according to Dartford (2021), the circular economy idea proposed in the EU is expected to have significant impacts on the relationship between the EU and its trading partners, more so with respect to the sustainability aspects of raw materials and products exported from and to Europe. To counter any negative impacts, the bottom line for any GND, including a universal one if the global player would reach that milestone would be to emphasize Green Growth, positing a strong approach to positive impacts even for less developed countries and Small Island Developing States (SIDS), observed to be struggling financially because of the debt crisis exacerbated by the impacts of COVID-19 (Allam, 2020b; United Nations, 2021).

The focus is on implementing policy frameworks that would encourage economies to continue investing in sustainable pathways, like as renewable energy and increased use of technologies in areas like urban planning. This, as noted by the World Bank (2021) will promote a gradual economic recovery, while at the same time ensuring that the global 2030 targets for reduction in emissions are not overlooked.

4.1 The GND as a Pathway for Climate Justice

Climate change represents complex and multifaceted challenges that have disproportionately impacted marginalized communities and economies across the globe. Due to climate change, a majority of the working class are in the current dispensation facing crises both of ecological nature and economic inequality. It has exposed them to health challenges, negatively impacted their quality of life, and promoted opportunities for unjust allocation of available resources. It is not surprising that approximately 1% of the global population is now earning almost two-thirds of the total wealth, generated globally each year. According to a report by Oxfam, their wealth if charged only 5% in the form of tax would help the world raise approximately $1.7 trillion annually; an amount enough to alleviate poverty for over 2 billion people (Oxfam, 2023). But unfortunately, such has not been the case; hence, prompting movements advocating for just and equality, with emerging as a potential pathway towards addressing these injustices, fostering equitable solutions, and establishing a just transition to a sustainable future. This section delves into the diverse role of the GND in achieving climate justice through the following three key dimensions below.

4.1.1 Addressing Historical Injustices and Empowering Marginalized Communities

From the literature, and discussion in previous chapters, it has been acknowledged that recognizing the historical context of colonialism, environmental racism, and unequal resource distribution is crucial for understanding the disproportionate burden of climate change faced by marginalized communities. This argument is not far-fetched as there is sufficient documented evidence that indigenous communities, minority groups, low-income populations, and SIDS often bear the brunt of climate impacts despite contributing minimally to emissions. For instance, a majority of SIDS are currently facing imminent danger on all fronts due to rising seawater levels, submersion of key infrastructures, and erratic weather conditions among many other challenges. These have not only hampered their ability to adapt but also to effectively garner enough resources to build resilience, improve their food security and protect their population from diverse issues. As of 2023, it is reported that over 370 million indigenous people have had their livelihoods destroyed by the impacts of climate change, while their health, traditional knowledge, and heritage is negatively impacted. It is important to note that the indigenous population though constitutionally only 6% of the global mass plays a significant role in the conservation of the world's biodiversity, and the risks posed by climate change do not only directly impact them but would have significant ramifications on global resources. It is reported that their effort benefits over 80% of the world biodiversity; hence, the need to take their plight into consideration (Redvers et al., 2023).

Identifying and respecting these vulnerable entities is critical, and need to start by inviting and acknowledging their value right at the discussion table. That is, they need to be included in all aspects of decision-making; whether political, economic, or social. Distribution of available resources, job allocation, and projects targeting climate action need to be equitably shared with those groups and economies receiving their rightful share.

Most of the current versions of GNDs from different countries emphasize the need for inclusive decision-making, with the plight of marginalized communities and vulnerable economies being considered. However, this needs to go beyond the planning and be reflected in the implementation phases of these GNDs. It should not be that such have been included just because of the voices of civil societies and independent organizations such as learning institutions. Divisions such as those witnessed in the United States between different political parties (Bradshaw et al., 2021) need to be addressed and people speak in unison towards supporting both the ideas, and their implementation, including in providing resources, and formulation of policies and regulations that would guarantee success and full implementation.

Such bipartisan support needs to be experienced in efforts focusing on tackling systemic inequalities embedded in energy, transportation, waste management systems, education, health sectors, and diverse economic sectors across the globe. Countries could borrow a leave from the United States where its version of GND received massive approval from the public; hence, the political class did not have much option but to respect the will and aspiration of the general populace (Boyle et al., 2021). In view of the advancing technological landscape, it would be critical to align the GNDs such that the aforementioned sectors are integrated to help create even more jobs, and streamline areas like financial services, making them available to everyone including those in marginalized areas. Technologies such as data analytics and modelling have the potential to provide evidence-based insights into the social, economic, and environmental impacts of the GND proposed; hence, allow for sound decision-making, and understanding of the potential benefits and trade-offs associated with the deal (Ahmed et al., 2022).

4.1.2 Promoting Equitable Solutions and a Just Transition

The second dimension through which GND can help achieve climate justice is through promoting equitable solutions. Among these is the 'Just Transition Framework' which prioritizes workers and communities most affected by the shift away from fossil fuels. The framework advocates for production and consumption approaches that are both holistic and with minimum or zero waste. Further, it calls for redress of past injustices through processes such as reparations, as well as forging new relationships that are anchored on shared prosperity and understanding. To ensure that workers are not discriminated against or disadvantaged, the proposed GNDs should emphasize the need for reskilling and upskilling to ensure decent work in green sectors (Wang & Lo, 2021). During the transition, there will be a need to provide social safety nets to support the vulnerable and also workers who might have been displaced or rendered jobless, and those who are affected by activities such as extraction and mining. As noted in a report by the International Labor Organization (ILO), the need for compensation and support to the vulnerable need not be negotiable, as most of the victims are from low-income households which mostly depend on primary sectors that are directly impacted by the transition, and often, those in this demographic may have hard times securing new jobs, or when successful, may face lower wages. Similarly, it has been noted that green technologies, including those that may favour the implementation of GNDs may tend to favour higher-skilled individuals, or people in certain geographical locations; hence, such need to be put into consideration (International Labor Organization and OECD, 2022).

4.1.3 Linking Global Climate Action with Local Struggles

The third dimension involves linking global climate action with local challenges and struggles. This can be achieved by focusing on a number of issues such as global solidarity, respect for local concerns, and focusing on locally proposed solutions that would have global impacts. Regarding global solidarity, the discussion in this paper has emphasized that GND recognizes climate change as a global challenge requiring international cooperation and solidarity. In all the diverse versions from different countries, there is a convergence in advocacy for collaborative efforts among nations, knowledge sharing, and technology transfer to ensure a just and effective global response.

While advocating for global solutions, the GND recognizes the importance of supporting local struggles and movements addressing climate injustices. This has not been the case in most countries or attempts to address climate change (Cheshmehzangi, 2020; Cheshmehzangi & Chen, 2021), as evidence shows that in most cases national priorities have always overshadowed the voices and concerns of local people, especially the marginalized ones (Allam et al., 2022b, 2023a, 2023b). For the GNDs to be considered effective, they must recognize the diverse experiences and priorities of communities worldwide, fostering partnerships between local and global actors, and amplifying the voices of those most affected (Levidow, 2022). The contribution of the local communities, as well as vulnerable economies toward climate change mitigation, job creation, and economic development must be acknowledged and supported in any form of GND. This then calls for a bottom-up approach to decision-making, rather than the top-down trend that has been dominant globally.

5 Conclusion

This concluding chapter, through the performance of bibliometric analysis, explores the theme of Green New Deals and maps out the theoretical conceptualizations, knowledge structure, and trends across different time scales and publications. This provides a deeper understanding of the evolution of the concept and can further provide researchers and policymakers with knowledge on how to better craft effective and contextually appropriate GND models. It is noted that GND adoption at the policy level is still in its infancy stage and is expected to keep gaining in popularity in the coming years, especially in view of the challenges brought about by the COVID-19 pandemic, underlining adequate and urgent economic responses aligning with national and international commitments to sustainability policies and agendas, including the Paris Agreement and the Agenda 2030 for Sustainable Development (Cheshmehzangi & Dawodu, 2019a, 2019b, 2019c). This subject is thus expected to gain in popularity among academic circles, which will lead to further popularization and adoption. The paper further underlines that while the subject of Green New Deal varies in interpretation, there is a strong convergence towards universally agreed principles, such as 'green growth' and 'green economy', hence showcasing foundational precepts to the concept, which can then be tailored to varied contexts, in accordance with their needs and agendas. Finally, it is noted the mapping of the foundational elements of varied GND models and frameworks could be an extremely interesting avenue for future research, to potentially lead towards a global GND framework and application roadmap, contextualized for implementation in different economies.

References

Ahmed, R., Shaheen, S., & Philbin, S. P. (2022). The Role of Big Data Analytics and Decision-Making in Achieving Project Success. *Journal of Engineering and Technology Management, 65*, 101697.

Allam, Z. (2020a). Chapter 7—Vital COVID-19 Economic Stimulus Packages Pose a Challenge for Long-Term Environmental Sustainability. In Z. Allam (Ed.), *Surveying the Covid-19 Pandemic and its Implications* (pp. 97–105). Elsevier.

Allam, Z. (2020b). Chapter 9—Oil, Health Equipment, and Trade: Revisiting Political Economy and International Relations During the COVID-19

Pandemic. In Z. Allam (Ed.), *Surveying the Covid-19 Pandemic and Its Implications* (pp. 119–127). Elsevier.

Allam, Z. (2020c). The Second 50 Days: A Detailed Chronological Timeline and Extensive Review of Literature Documenting the COVID-19 Pandemic from Day 50 to Day 100. *Surveying the COVID-19 Pandemic Its Implications*, 9.

Allam, Z. (2020d). *Surveying the Covid-19 Pandemic and Its Implications: Urban Health, Data Technology and Political Economy*. Elsevier Science.

Allam, Z. (2020e). The Third 50 Days: A Detailed Chronological Timeline and Extensive Review of Literature Documenting the COVID-19 Pandemic From Day 100 to Day 150. *Surveying the Covid-19 Pandemic its Implications*, 41.

Allam, Z. (2020f). *Urban Governance and Smart City Planning: Lessons from Singapore*. Emerald Group Publishing.

Allam, Z., Chabaud, D., Gall, C., et al. (2023a). Chapter 8—Financing the 15-Minute City Concept and Its Infrastructural Ecosystem in Developing Nations Through Fiscal Mechanisms. In Z. Allam, D. Chabaud, C. Gall, et al. (Eds.), *Resilient and Sustainable Cities* (pp. 121–133). Elsevier.

Allam, Z., Cheshmehzangi, A., & Jones, D. S. (2023b). Redefining Climate Policy and Economic Resilience in the Consuming World. In *Climate and Social Justice: The Political Economy of Urban Resilience and Mercantilism* (pp. 77–102). Springer Nature Singapore.

Allam, Z., & Jones, D. S. (2020). Pandemic Stricken Cities on Lockdown. Where Are Our Planning and Design Professionals [Now, Then and Into the Future]? *Land Use Policy, 97*, 104805.

Allam, Z., & Jones, D. S. (2021). Future (Post-COVID) Digital, Smart and Sustainable Cities in the Wake of 6G: Digital Twins, Immersive Realities and New Urban Economies. *Land Use Policy, 101*, 105201.

Allam, Z., Jones, D., Biyik, C., et al. (2021a). Rewriting City Narratives and Spirit: Post-Pandemic Urban Recovery Mechanisms in the Shadow of the Global 'Black Lives Matter' Movement. *Research in Globalization, 3*, 100064.

Allam, Z., & Newman, P. (2018). Redefining the Smart City: Culture, Metabolism and Governance. *Smart Cities, 1*, 4–25.

Allam, Z., & Newman, P. (2023). Smart Cities and the Urban Economy: How Smart City Can Integrate Good Urban Economies. In *Revising Smart Cities with Regenerative Design* (pp. 101–116). Springer International Publishing.

Allam, Z., Sharifi, A., Bibri, S. E., et al. (2022a). The Metaverse as a Virtual Form of Smart Cities: Opportunities and Challenges for Environmental, Economic, and Social Sustainability in Urban Futures. *Smart Cities, 5*, 771–801.

Allam, Z., Sharifi, A., Giurco, D., et al. (2021b). On the Theoretical Conceptualisations, Knowledge Structures and Trends of Green New Deals. *Sustainability, 13*, 12529.

Allam, Z., Sharifi, A., Giurco, D., et al. (2022b). Green New Deals Could Be the Answer to COP26's Deep Decarbonisation Needs. *Sustainable Horizons, 1*, 100006.

Boyle, A. D., Leggat, G., Morikawa, L., et al. (2021). Green New Deal Proposals: Comparing Emerging Transformational Climate Policies at Multiple Scales. *Energy Research & Social Science, 81*, 102259.

Bradshaw, C. J. A., Ehrlich, P. R., Beattie, A., et al. (2021) Underestimating the Challenges of Avoiding a Ghastly Future. *Frontiers in Conservation Science, 1*, 615419.

Brian, K., & Patrick, L. (2010). *OECD Insights From Crisis to Recovery The Causes, Course and Consequences of the Great Recession: The Causes, Course and Consequences of the Great Recession*. OECD Publishing.

Chabaud, D., Allam, Z., Gall, C., et al. (2023). Redefining Investable Infrastructure in Developing Nations in a Postpandemic Era: The Case of the 15-Minute City. In *Resilient and Sustainable Cities* (pp. 136–149). Elsevier.

Chamoso, P., Gonzalez-Briones, A., Rodríguez, S., et al. (2018). Tendencies of Technologies and Platforms in Smart Cities: A State-of-the-Art Review. *Wireless Communications and Mobile Computing*, 1–17.

Cheshmehzangi, A. (2020). The Analysis of Global Warming Patterns from 1970s to 2010s. *Atmospheric and Climate Sciences, 10*(3), 392–404. https://doi.org/10.4236/acs.2020.103022

Cheshmehzangi, A., & Chen, H. (2021). *China's Sustainability Transitions: Low Carbon and Climate-Resilient Plan for Carbon Neutral 2060*. Springer.

Cheshmehzangi, A., & Dawodu, A. (2019a). *Sustainable Urban Development in the Age of Climate Change—People: The Cure or Curse*. Palgrave Macmillan.

Cheshmehzangi, A., & Dawodu, A. (2019b). Stakeholder Methods to Climate Change and Sustainable Development. In *Sustainable Urban Development in the Age of Climate Change—People: The Cure or Curse* (pp. 23–68). Palgrave Macmillan.

Cheshmehzangi, A., & Dawodu, A. (2019c). Case Study Reviews: People, Perspective, and Planning. In *Sustainable Urban Development in the Age of Climate Change—People: The Cure or Curse* (pp. 69–131). Palgrave Macmillan.

Cheshmehzangi, A., Dawodu, A., & Sharifi, A. (2021). *Sustainable Urbanism in China*. Routledge.

Cheshmehzangi, A., & Tang, T. (2022). *China's City Cluster Development in the Race to Carbon Neutrality*. Springer.

Cheshmehzangi, A., You, N., Siri, J., & Birch, E. (2024). *Harnessing Urban Innovation to Unlock the Sustainable Development Goals*. Springer.

Congress. (2019). 116th Congress 1st Session: H. RES. 109. In Representative Ho (Ed.). Congress.

Congressional Research Service. (2019). *China's Economic Rise: History, Trends, Challenges, and Implications for the United States*. CRS Report.

Conte, N. (2019). *Green New Deal*. https://wtpartnership.co/green-new-deal/

Dartford, K. (2021). *How can the European Green Deal Avoid 'Potential Frictions' with Other Countries?* https://www.euronews.com/green/2021/02/03/how-can-the-european-green-deal-avoid-potential-frictions-with-other-countries

Eurofound. (2023). *Economic and Social Inequalities in Europe in the Aftermath of the COVID-19 Pandemic*. Publications Office of the European Union.

European Commission. (2019). *The European Green Deal: Questions & Answers*. https://ec.europa.eu/commission/presscorner/detail/en/QANDA_19_6690

European Union External Action. (2023). *Europe Knows That Climate Action Is Vital to Global Security*. https://www.eeas.europa.eu/eeas/europe-knows-climate-action-vital-global-security_en

Evans, D. (2014). The History and Evolution of the Affordable Housing Design Advisor: Bringing the Power of Design to Affordable Housing. *Cityscape, 16*, 87–102.

Frank, M. R., Sun, L., Cebrian, M., et al. (2018). Small Cities Face Greater Impact from Automation. *Journal of Royal Society, Interface, 15*, 20170946.

Fridays for Future. (2021). *Who We Are*. https://fridaysforfuture.org/what-we-do/who-we-are/

Gascó-Hernandez, M. (2018). Building a Smart City: Lessons from Barcelona. *Communications of the ACM, 61*, 50–57.

González-Sánchez, M., & Martín-Ortega, J. L. (2020). Greenhouse Gas Emissions Growth in Europe: A Comparative Analysis of Determinants. *Sustainability, 12*, 1012.

Gopinath, G. (2021). *World Economic Outlook Update: Policy Support and Vaccines Expected to Lift Activity*. https://www.imf.org/en/Publications/WEO/Issues/2021/01/26/2021-world-economic-outlook-update

Intergovernmental Panel on Climate Change. (2014). Impacts, Adaptation, and Vulnerability. In C. B. Field, V. R. Barros, D. J. Dokken, et al. (Eds.), *Contribution of Working Group II to the Fifth Assessment Report of the Intergovernmental Panel on Climate Change*. IPCC.

International Labor Organization and OECD. (2022). *Promoting a Just and Inclusive Green Transition*. ILO-OECD.

IPCC. (2021). The Physical Science Basis. In V. Masson-Delmotte, P. Zhai, A. Pirani, et al. (Eds.), *Contribution of Working Group I to the Sixth Assessment Report of the Intergovernmental Panel on Climate Change*. Cambridge University Press.

Joskow, P. L. (2001). *U.S. Energy Policy During the 1990s*. https://economics.mit.edu/files/1144

Kolotouchkina, O., & Seisdedos, G. (2017). Place Branding Strategies in the Context of New Smart Cities: Songdo IBD, Masdar and Skolkovo. *Place Branding and Public Diplomacy, 14*, 115–124.

Levidow, L. (2022). Green New Deals: What Shapes Green and Deal? *Capitalism Nature Socialism, 33*, 76–97.

London Assembly. (2021). *A Green New Deal*. https://www.london.gov.uk/coronavirus/londons-recovery-coronavirus-crisis/recovery-context/green-new-deal

Marcacci, S. (2020). *Renewable Energy Prices Hit Record Lows: How Can Utilities Benefit from Unstoppable Solar and Wind?* https://www.forbes.com/sites/energyinnovation/2020/01/21/renewable-energy-prices-hit-record-lows-how-can-utilities-benefit-from-unstoppable-solar-and-wind/?sh=29a8522d2c84

O'Neill, A. (2023). *South Korea: Growth of Real Gross Domestic Product (GDP) from 1954 to 2022, with Projections Until 2028*. https://www.statista.com/statistics/263758/gross-domestic-product-gdp-growth-in-south-korea/

Oxfam. (2023). *Richest 1% Bag Nearly Twice as Much Wealth as the Rest of the World Put Together Over the Past Two Years*. https://www.oxfam.org/en/press-releases/richest-1-bag-nearly-twice-much-wealth-rest-world-put-together-over-past-two-years

Redvers, N., Aubrey, P., Celidwen, Y., et al. (2023). Indigenous Peoples: Traditional Knowledges, Climate Change, and Health. *PLOS Glob Public Health, 3*, e0002474.

Rohaidi, N. (2018). *Singapore Wins Smart City of 2018 Award*. https://govinsider.asia/smart-gov/singapore-wins-smart-city-of-2018-award/

Roser, M., Ritchie, H., & Ortiz-Ospina, E. (2019). *World Population Growth*. https://ourworldindata.org/world-population-growth#citation

Samad, M. E. S., Hawken, S., Sargolzaei, S., et al. (2019). Implementing Citizen Centric Technology in Developing Smart Cities: A Model for Predicting the Acceptance of Urban Technologies. *Technological Forecasting and Social Change, 142*, 105–116.

The World Bank. (2022). Reconsidering Sites and Services: A Global Review. *The World Bank Review, 1*, 301–335.

To, W.-M., Lee, P. K. C., & Lam, K.-H. (2018). Building Professionals' Intention to Use Smart and Sustainable Building Technologies—An Empirical Study. *PLoS ONE, 13*, e0201625.

Tollefson, J. (2021). COVID-19 Curbed Carbon Emissions in 2020—But Not by Much. *Nature, 589*, 343.

UNEP. (2014). *Emerging Issues for Small Island Developing States*. UNEP.

UNFCCC. *Vulnerability and Adaptation to Climate Change in Small Island Developing States*. https://unfccc.int/files/adaptation/adverse_effects_and_response_measures_art_48/application/pdf/200702_sids_adaptation_bg.pdf

United Nations. (2017). New Urban Agenda. *Habitat, III*, 1–66.

United Nations. (2021). *Small Island Developing States Call for Enhanced Support for a Resilient Post COVID-19 Recovery*. https://www.un.org/ohrlls/news/small-island-developing-states-call-enhanced-support-resilient-post-covid-19-recovery

United Nations Framework Convention on Climate Change. (1998). *Kyoto Protocol to the United Nations Framework Convention on Climate Change*. https://unfccc.int/resource/docs/convkp/kpeng.pdf

United Nations Framework Convention on Climate Change. (2015). *Paris Agreement*. https://unfccc.int/sites/default/files/english_paris_agreement.pdf

Walker, R. A. (2022). From the New Deal to the Green New Deal. In C. Calhoun & B. Y. Fong (Eds.), *The Green New Deal and the Future of Work* (pp. 23–52). Columbia University Press.

Wang, X., & Lo, K. (2021). Just Transition: A Conceptual Review. *Energy Research & Social Science, 82*, 102291.

World Bank. (2021). *COVID-19 to Plunge Global Economy into Worst Recession Since World War II*. https://www.worldbank.org/en/news/press-release/2020/06/08/covid-19-to-plunge-global-economy-into-worst-recession-since-world-war-ii

Index

Z. Allam and A. Cheshmehzangi, *Sustainable Futures and Green New Deals*, https://doi.org/10.1007/978-3-031-63642-4

GPSR Compliance

The European Union's (EU) General Product Safety Regulation (GPSR) is a set of rules that requires consumer products to be safe and our obligations to ensure this.

If you have any concerns about our products, you can contact us on ProductSafety@springernature.com

In case Publisher is established outside the EU, the EU authorized representative is:

Springer Nature Customer Service Center GmbH
Europaplatz 3
69115 Heidelberg, Germany